154
Advances in Biochemical Engineering/Biotechnology

Aims and Scope

This book series reviews current trends in modern biotechnology and biochemical engineering. Its aim is to cover all aspects of these interdisciplinary disciplines, where knowledge, methods and expertise are required from chemistry, biochemistry, microbiology, molecular biology, chemical engineering and computer science.

Volumes are organized topically and provide a comprehensive discussion of developments in the field over the past 3–5 years. The series also discusses new discoveries and applications. Special volumes are dedicated to selected topics which focus on new biotechnological products and new processes for their synthesis and purification.

In general, volumes are edited by well-known guest editors. The series editor and publisher will, however, always be pleased to receive suggestions and supplementary information. Manuscripts are accepted in English.

In references, Advances in Biochemical Engineering/Biotechnology is abbreviated as *Adv. Biochem. Engin./Biotechnol.* and cited as a journal.

More information about this series at http://www.springer.com/series/10

Gérald Thouand · Robert Marks
Editors

Bioluminescence: Fundamentals and Applications in Biotechnology - Volume 3

With contributions by

M.M. Calabretta · D. Calabria · L. Cevenini
Pimchai Chaiyen · M. Cregut · Sylvia Daunert
Emre Dikici · M.J. Durand · Kristen Grinstead
A. Hua · Rodica Elena Ionescu · Kun Jia · Smita Joel
S. Jouanneau · Issmat I. Kassem · Leslie D. Knecht
A. Lahmar · E. Michelini · Sally Miller · Gregory O'Connor
Patrizia Pasini · Gireesh Rajashekara · A. Roda
Nelson Salgado · Gary A. Splitter · Sebastian Strobel
G. Thouand · Ruchanok Tinikul · Jean-Marc Zingg

Springer

Editors
Gérald Thouand
UMR CNRS 6144
Universite de Nantes
Nantes
France

Robert Marks
Department of Biotechnology Engineering
Ben-Gurion University of the Negev
Beersheba
Israel

ISSN 0724-6145 ISSN 1616-8542 (electronic)
Advances in Biochemical Engineering/Biotechnology
ISBN 978-3-319-27405-8 ISBN 978-3-319-27407-2 (eBook)
DOI 10.1007/978-3-319-27407-2

Library of Congress Control Number: 2014943751

Printed on acid-free paper

This Springer imprint is published by SpringerNature
The registered company is Springer International Publishing AG Switzerland

Preface

In the wake of the International Year of light launched by the UNESCO in 2015, this book (the third of a 3 volumes), brings to readership new insights of the bioluminescence phenomenon and applications. Admired since the Antiquity, the biochemical reaction was studied during the last century and remains open to new discoveries and many applications were done with this bright light. Thanks to this book, readers will not understand only the magic of the reaction but also the abundance of applications in all major sectors (environment, food and health).

Contents

Part I
Fundamentals of Bioluminescence

Adv Biochem Eng Biotechnol (2016) 154: 3–18
DOI 10.1007/10_2015_325

Published Online: 22 April 2015

Luciferase Genes as Reporter Reactions: How to Use Them in Molecular Biology?

L. Cevenini, M.M. Calabretta, D. Calabria, A. Roda and E. Michelini

Abstract The latest advances in molecular biology have made available several biotechnological tools that take advantage of the high detectability and quantum efficiency of bioluminescence (BL), with an ever-increasing number of novel applications in environmental, pharmaceutical, food, and forensic fields. Indeed, BL proteins are being used to develop ultrasensitive binding assays and cell-based assays, thanks to their high detectability and to the availability of highly sensitive BL instruments. The appealing aspect of molecular biology tools relying on BL reactions is their general applicability in both in vitro assays, such as cell cultures or purified proteins, and in vivo settings, such as in whole-animal BL imaging. The aim of this chapter is to provide the reader with an overview of state-of-the-art bioluminescent tools based on luciferase genes, highlighting molecular biology strategies that have been applied so far, together with some selected examples.

Keywords Bioluminescence · Molecular biology tools · Luciferase · Reporter gene · In vivo bioluminescence imaging

Abbreviations

BL	Bioluminescence
PpyLuc	North American *Photinus pyralis* firefly luciferase
BRET	Bioluminescence Resonance Energy Transfer
PCA	Protein Complementation Assay
GPCR	G protein coupled receptors
CCD	Charge-coupled device
CMOS	Complementary metal-oxide semiconductors
HTS	High Throughput Screening

L. Cevenini · M.M. Calabretta · D. Calabria · A. Roda · E. Michelini (✉)
Dept. of Chemistry "G. Ciamician", University of Bologna, Via Selmi 2, 40126 Bologna, Italy
e-mail: elisa.michelini8@unibo.it

Contents

1 Introduction

Bioluminescence (BL) is the light emission produced as a consequence of a chemical reaction occurring in a living organism [1]. The intensity of the emitted light depends upon the overall efficiency of the bioluminescent reaction (Φ_{BL}), calculated as the ratio between emitted photons and reacting substrate molecules. This reaction leads to electronically excited products that either decay emitting photons of visible light or act as sensitizers, passing their energy to another chemical species that in turn will emit light. Different from fluorescence, background levels in bioluminescence measurements are extremely low, resulting in very high signal-to-noise ratios [2, 3]. The amazing properties of bioluminescent systems make them powerful bioanalytical tools for tracking molecules, cells, and even protein–protein interactions, both in vitro and in vivo settings. It goes without saying that BL has endless applications in several fields, including pharmaceutical, environmental, forensic, and food analysis [4–6].

Since the cloning of the first luciferase in the 1970s [7], several other bioluminescent systems have been investigated and a wide portfolio of bioluminescent tools is currently available. In fact, numerous bioluminescent enzymes, known as luciferases, were isolated from luminescent organisms, including several firefly species, bacteria, and marine organisms, and the corresponding genes were cloned [8–10]. In addition, a number of luciferases with altered emission properties (e.g. shifted emission wavelengths, higher quantum yield emission, longer kinetics) were obtained by random or site-directed mutagenesis. Indeed, as for the green fluorescent proteins, a palette of different luciferases relying on diverse biochemical reactions and requiring different BL substrates is now commercially available, together with tailored substrates [11]. Nevertheless, research in this field is very active and new luciferases are being investigated, guided by theoretical studies unravelling the molecular mechanisms of the BL reaction [12].

The typical goal of BL measurements involves the study of gene expression and gene regulation, as well as the detection of target analytes present in a sample. Thanks to their high detectability and to the availability of highly sensitive BL instruments, BL proteins have been successfully used to develop ultrasensitive binding assays and cell-based assays. Interestingly, molecular biology tools based on BL reactions are highly versatile, being suitable for both in vitro assays, such as cell cultures or purified proteins, and in vivo settings, such as in whole-animal BL imaging.

The aim of this chapter is to provide the reader with an overview of state-of-the-art BL tools based on luciferase genes, highlighting molecular biology strategies that have been applied so far, together with some selected examples.

2 Luciferases: Choosing the Right One

BL proteins are generally referred as luciferases, which are enzymes catalyzing chemical reactions involving the oxidation of a substrate, generically known as luciferin. The reaction yields a single-excited intermediate that decays with photon emission. Since the enzyme-active site provides a reaction microenvironment that is favorable to radiative decay of the excited intermediate product, BL reactions are characterized by very high quantum yields (0.44 for the reaction of North American *Photinus pyralis* firefly luciferase, PpyLuc) [2].

Among luciferases, PpyLuc is by far the most investigated and employed in bioanalytical assays. PpyLuc is a 61-kDa monomeric protein that does not require any post-translational modifications and it does not show any toxicity to cells, even at high concentrations; thus, it is suitable for heterologous expression in both prokaryotic and eukaryotic systems. It catalyzes the oxidation of the substrate D-luciferin in the presence of adenosine triphosphate (ATP) and oxygen, yielding emission of yellow green light ($\lambda_{max} = 560$ nm at pH 7.8). PpyLuc bioluminescence is temperature and pH sensitive, showing a remarkable red shift at lower pH and higher temperatures [13]. The half-life of this luciferase expressed in mammalian cells was calculated to be in the range from 1 to 4 h in the presence of protein inhibitors [14, 15]. When used in drug screening assays, several groups reported that this luciferase is inhibited by small molecules [16, 17]. Ho et al. [18] screened a library of 42,460 chemicals (at fixed 10 μM concentration) with different reporter genes (among which RLuc, β-lactamase, *R. reniformis* luciferase variants, a *Luciola cruciata* luciferase variant, a *Gaussia princeps* mutant, and NanoLuc), reporting that hit rates ranged from less than 0.1% for β-lactamase to up to 10% for Renilla luciferase variants with high degrees of inhibitor overlap (40–70 %) obtained with related luciferases. Therefore, optimization of assays in which two structurally diverse luciferases are co-expressed is preferred to avoid false positive in high-throughput screening assays.

Interestingly, a number of firefly luciferase mutants have been obtained by random and site-directed mutagenesis, with a palette of luciferase variants emitting in the range of 535–620 nm and with different properties (e.g., thermostability, pH-independence emission, destabilized proteins) [19, 20].

There is also a variety of bacterial luciferases, among which the most studied is the luciferase-luciferin (lux) operon from *Vibrio fischeri*. This luciferase, which is a flavin-dependent monooxygenase, has been deeply described by Dunlap et al. [21]. Briefly, it uses reduced flavin, oxygen, and long-chain aldehyde as substrates, yielding oxidized flavin, carboxylic acid, and H_2O as products. This reaction is accompanied by emission of blue-green light (λ_{max} = 490 nm).

The majority of marine luciferases contain a natural signal peptide for secretion that allows BL measurements to be performed directly in the medium, without the need for destroying cells or tissues [22]. The use of secreted luciferases is receiving increased attention thanks to the possibility to perform longitudinal analysis with repetitive measurements of the same sample. Unlike the firefly luciferase systems, these coelenterazine-utilizing luciferases do not require accessory high-energy molecules such as ATP, thus simplifying assay conditions.

Gaussia luciferase (Gluc), cloned from the copepod *Gaussia princeps,* which catalyzes the oxidation of the substrate coelenterazine in a reaction that emits light (λ_{em} = 470 nm) and Cypridina luciferase, cloned from the ostracod *Cypridina noctiluca* (λ_{em} = 465 nm), have been widely used in single- and dual-reporter assays and for ex vivo real-time monitoring of in vivo biological processes [23, 24]. It must be pointed out that, despite these luciferases being highly advantageous thanks to their small size and ATP independence, coelenterazine and other substrates for marine luciferases (e.g., furimazine) are more expensive, less stable, and self-luminescent, producing high signal background.

NanoLuc luciferase (Nluc) is a newly engineered luciferase that catalyses the oxidation of a novel imidazopyrazinone substrate (furimazine) to produce light [25]. This luciferase appears to be the most sensitive reporter, leading to ultra-sensitive reporter gene assays [26–28]. Loh et al. engineered group A Streptococcus (GAS) strains with Ppyluc or Nanoluc and reported that during the logarithmic phase Nluc-labeled bacteria emitted a BL signal up to 15 times higher than Ppyluc-labeled bacteria [29]. Even if the reaction catalysed by Nanoluc does not require ATP, using a permeant pro-substrate that is reduced only by living cells, it is still possible to use this reporter to monitor cell viability in real time (Table 2).

Other secreted luciferases, such as Metridia and *Cypridina noctiluca* luciferase, have been cloned in recent years, but they apparently show less sensitivity than Nanoluc [9, 30]. Interestingly, *Metridia longa* luciferase isoform MLuc7, with its size of 16.5 kDa, is the smallest natural luciferase cloned to date; it was demonstrated to be an efficient reporter in mammalian cells [31].

Thanks to the advances in molecular biology and from naturalists exploring at remote areas of the planet at night, the field is continuously growing and new BL systems will be discovered. Tables 1 and 2 list some luciferases and BL substrates that are currently commercially available (although these lists are far from being exhaustive). Several commercially available luciferase assay formulations have been developed to measure the reporter activity in single-step, add, and read assays.

3 Reporter Gene Assays

Luciferases are generally considered as reporter genes (i.e., genes whose expression is easily and quantitatively detected), which can be attached to a regulatory sequence of a target gene for investigating its regulation or intra-cellular location [32, 33].

Table 1 Commercially available luciferases whose coding sequence is for reporter gene applications, calcium detection, and in vivo imaging

Luciferases	Organism	Length (aa)	Size (MW, KDa)	Substrate	BL emission λ_{max} (nm)	Company
Ppyluc[a]	*P. pyralis*	550	61	D-luciferin	557	Several companies
Gaussia luciferase[a]	*G. princeps*	185	19.9	Coelenterazine	482	ThermoScientific, Targeting Systems
Renilla luciferase[a]	*R. reniformis*	312	36	Coelenterazine	475	Promega, ThermoScientific, Targeting Systems
Nanoluc[a]	*Oplophorus gracilirostris*	171	19.1	Furimazine	465	Promega
ELuca (emerald)	*Pyrearinus termitilluminans*	543	60.7	D-luciferin	538	Toyobo (Osaka, Japan)
Lucia	n.r.	n.r.	23	Coelenterazine	n.r.	InvivoGen
Red-emitting Luciola luciferase	*L. italica*	548	61	D-luciferin	610	Targeting systems
CBG99	*P. plagiophthalamus*	542	64	D-luciferin	537	Promega
CBR	*P. plagiophthalamus*	542	64	D-luciferin	613	Promega
Cypridina luciferase	*Cypridina noctiluca*	555	62	Cypridina luciferin	463	Prolume RTD, Targeting systems

[a]Luciferases available in different forms (e.g., intracellular, secreted, and shorter-lived forms)
n.r. not reported

Table 2 Commercial substrates for luciferases: preferred applications and producers

Commercial name	Description/preferred application	Company
RediJect D-luciferin	In vivo imaging	PerkinElmer
Britelite plus, Neolite, Steadylite	Cell-based assays	PerkinElmer
BrightGlo, OneGlo, SteadyGlo, ChromaGlo	Cell-based assays	Promega
VivoGlo Z-DEVD-Aminoluciferin sodium salt	In vivo applications	Promega
NanoGlo	Cell-based assays	Promega
Ready-to-use D-luciferin substrate	BL imaging	System Biosciences
Coelenterazine native and derivatives (*cp*, *f*, *fcp*, *h*, *hcp*, *i*, *ip*, *n*)	In vitro and in vivo applications	Promega, NanoLight Life Technologies, Biotium, Perkin Elmer
D-luciferin free acid and sodium/potassium salts	General use	Promega, Sigma-Aldrich Synchem Regis Technologies
DMNPE-caged D-luciferin	D-luciferin ester derivative which is continuously hydrolyzed inside the cell to supply D-luciferin	Biotium, Invitrogen
Dual-luciferase reporter assay system	Dual-reporter assays	Promega
Pierce cypridina-firefly luciferase dual assay kit (vargulin and D-luciferin) Pierce gaussia-firefly luciferase dual assay kit	Dual-reporter assays	ThermoScientific
LiveResponse luciferase assay reagents	Cell-based HTS applications	Targeting System
SUMO2-aminoluciferin	HTS for modulators of dSUMOylase activity	LifeSensors
RealTime-Gl MT cell viability assay	Non lytic cell viability assays	Promega

Cells (bacteria, yeasts, or mammalian cells) can be genetically modified by introducing a luciferase, used as a reporter gene, fused to a regulatory DNA sequence that is activated only in the presence of a target analyte (e.g., genotoxic compounds, metals, hormones, xenobiotics), thus regulating the reporter gene expression [34]. It is worthwhile to note that such cell-based systems measure the bioavailable fraction of the analyte, which is crucial information that is difficult to obtain with other analytical techniques [35, 36]. By using living systems as sensing elements, it is possible to measure different physiological responses, including membrane damage, apoptosis, necrosis, and alterations of signal transduction or second messenger pathways [37].

Continuous advances in molecular biology and new techniques based on massive sequencing have allowed the identification of genes and cellular pathways that constitute potential therapeutic targets [38]. In the past few years, there has been a veritable explosion in the field of reporter genes, and BL has been used as the basic tool to assess the ability of candidate drugs to interact with such target genes and cellular pathways. When compared to binding assays and in vitro biochemical assays, cell-based assays provide a more physiological environment and thus give information about the actual effect of a compound, providing direct evidence of its activity.

To date, a variety of BL cell-based assays and configurations have been reported. Figure 1 shows some examples of genetic constructs that can be used in reporter gene technology. Luciferase can be put under the regulation of a constitutive promoter, a tissue specific promoter, or an inducible promoter. Different luciferases can also be expressed in the same genetic construct, either as fusion proteins or linked by an internal ribosome entry site sequence or ribosome skipping (2A sequence), which allows for translation of either the target gene and the reporter or two reporters simultaneously [39].

Two major directions can be envisaged: the use of two or more reporter proteins in the same cell to achieve multiplexing or the integration of the cells into portable analytical devices for on-site applications [38, 40].

Concerning the first trend, Fig. 2 shows the different strategies that can be selected for developing a multiplex cell-based assay relying on two different luciferases expressed in the same cell. Basically, the resolution of the two BL signals emitted by the different luciferases can be achieved by exploiting spatial resolution (i.e., by expressing an intracellular and a secreted luciferase and

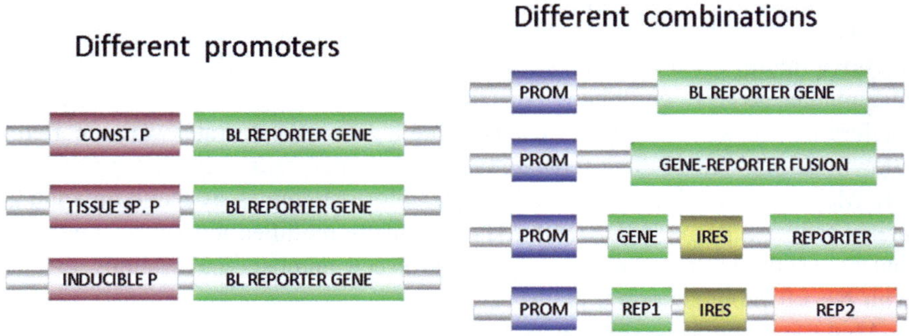

Fig. 1 Reporter gene constructs. Different configurations can be exploited by placing the reporter gene under the control of a constitutive promoter (const. p), a promoter with tissue specific activation (tissue sp. p), or an inducible promoter (inducible p). Different combinations of promoters can be used to express two genes (target gene and reporter gene or two reporter genes) as a fusion protein or separated by internal ribosome entry site sequences

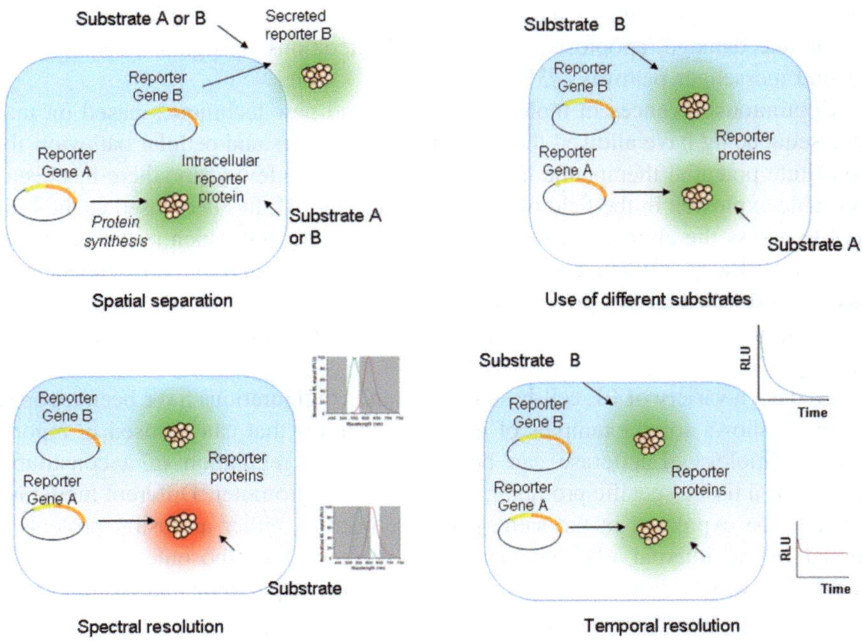

Fig. 2 The different strategies that can be exploited for developing a multiplex cell-based assay relying on two different luciferases expressed in the same cell. See text for details

measuring the two signals in the cells or in the medium); temporal resolution (i.e., by expressing a luciferase with flash-type emission and another with glow-type emission); spectral resolution (i.e., by exploiting two luciferase emitting at a different wavelength), or by simply performing sequential measurements of BL emission by two luciferases requiring different substrates [41–43].

The recent availability of new reporter genes with improved BL properties, together with technical advancements, prompted the development of robust and sensitive multiplexed and multicolour cell-based assays. Dual and triple-color mammalian assays, combining spectral unmixing of green- and red-emitting luciferases with secreted luciferases requiring a different substrate, were reported to allow measurement of three separate targets with high sensitivity and rapidity [41, 44]. Mezzanotte et al. [45] reported a triple-color imaging assay based on a human breast cancer cell line (MDA-MB-231) genetically modified to express green, red, and blue light emitting luciferases to monitor cell numbers and viability together with NF-κB promoter activity.

Multiplex reporter systems have been developed to monitor multiple biological variables in real time. A multiplex reporter assay was reported to rely on 10 different tags fused to secreted *Gaussia* luciferase, allowing the real-time monitoring of tumor subpopulations in vitro and in vivo by immunoblotting of the tags [46].

In relation to the development of portable systems relying on BL cells, a lot of effort was also aimed at integrating both the biorecognition element (the cell) and the transducer (detector such as CMOS and CCD) in compact devices, possibly for online detection [47–50].

Despite their appealing features, conventional reporter assays do not capture dynamic processes and the intrinsic variability among cells. To provide such information, cellular bioluminescence imaging is required, providing advantageous properties when compared to fluorescence imaging, such as no photobleaching, no autofluorescence, and no phototoxicity [51].

Dual-color imaging was developed based on a green-emitting luciferase from a Brazilian click beetle (ELuc) and a red-emitting luciferase from a railroad worm (SLR). ELUc and SLR were localized to the peroxisome and the nucleus under the regulation of a CMV and NF-κB responsive elements. Imaging at the subcellular level was obtained using a CellGraph (ATTO) luminescence microscope at 37 °C equipped with short-pass and long-pass filters [52].

4 Protein–Protein Interaction BL Assays

Because of their ability to unravel protein–protein interactions, bioluminescence resonance energy transfer (BRET) and split complementation strategies are increasingly gaining attention. BRET is a resonance energy transfer process occurring between a light-emitting luciferase donor and an acceptor fluorescent protein [53–55]. When the donor and the acceptor are brought into close proximity to one another (1–10 nm) and are properly oriented, the former transfers its energy to the latter in a nonradiative way. Because the acceptor emits light at a different wavelength than the donor, the energy transfer can be easily detected by measuring the ratio of the acceptor emission intensity to the donor emission intensity (Fig. 3a). The signal of a BRET assay is ratiometric, so well-to-well differences due to cell number or signal decay across a plate are compensated. Because BRET strictly depends on the molecular proximity between donor and acceptor, it is suitable for monitoring the activation state of any receptor that undergoes association or conformational changes after ligand binding. For example, to evaluate receptor dimerization, the two subunits of the receptor are genetically tagged either with the BL donor (e.g. Renilla or firefly luciferase) or the acceptor (e.g. a green fluorescent protein variant).

If the activation of the receptor brings the donor and acceptor in a favorable position, BRET will occur. This phenomenon can be either observed in vitro, using purified proteins, or within the cell where the fusion proteins were produced. Several studies reported the superiority of BRET over fluorescence resonance energy transfer (FRET) strategy; for example, a genetically encoded maltose biosensor was developed based on maltose binding protein (MBP) flanked by a Renilla luciferase variant (RLuc2) at the C-terminus and a green fluorescent protein (GFP (2)) at the N-terminus. It showed a 30% increase in the BRET signal compared to a

Fig. 3 **a** Schematic view of BRET-based assay in living cells and **b** schematic representation of protein fragment complementation assay. Reprinted with permission from Ref. [73]

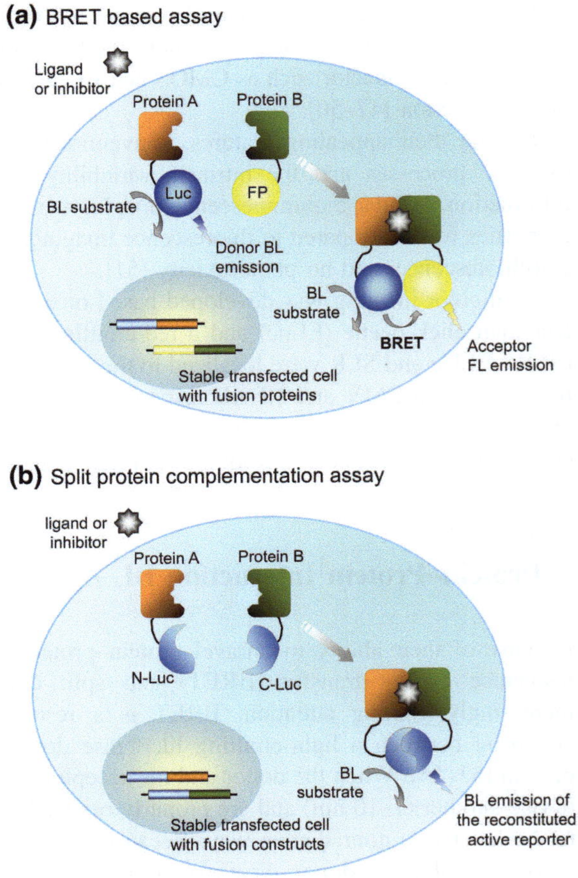

10% increase with an analogous FRET-based sensor upon binding of the same maltose concentration [56].

Conversely, split complementation strategies, also called protein–fragment complementation assays (PCAs), are based on genetic fusions of hypothetical binding partners to two rationally designed fragments of a reporter protein. The interaction of the two target partners allows for reconstitution of a functionally active reporter protein [57, 58] (Fig. 3b). PCAs have been widely applied for monitoring protein–protein interactions in vitro and in vivo, enabling the detection of even weak protein interactions. Moreover, such approaches are suitable for high-throughput analysis. One major drawback of PCA is the possibility of spontaneous reconstitution of dissected fragments, even in absence of the binding proteins, leading to false positives. The choice of the optimum dissection sites is a crucial issue and must be carefully optimized according to the protein–protein interactions under investigation [59]. In addition, in order to develop sensitive assays, it is usually required to overexpress the fusion proteins, leading to a nonphysiological intracellular environment. In this context, the availability of more active luciferases could allow one to design these strategies by expressing genetic constructs at physiological level, resulting in more predictive assays.

Both BRET and PCA have been widely used for monitoring the activation of G protein-coupled receptors (GPCRs), cytoplasmic Myc protein (c-Myc), which is deregulated in several cancers [60], and nuclear steroid receptors [61]. Interestingly, BRET also has been applied to photodynamic therapy to avoid the need for an external excitation light source, which represents an issue for clinical applications due to limitations of tissue-penetrating properties [62].

5 In Vivo BL Imaging

A powerful approach for monitoring molecular targets and biochemical processes in real time within living animals is represented by BL whole-animal molecular imaging [63–65]. BL imaging represents a sensitive, quantitative, and real-time method to monitor gene expression patterns, tumor growth and metastasis, infections, and stem cell homing in living animals. These in vivo assays provide highly predictive pre-clinical data in accordance with the 3Rs principle—replacement, reduction, and refinement—in drug discovery [66]. BL imaging can efficiently increase the amount (and improve the quality) of quantitative data collected. Continuous imaging of a specific process in a given animal subject reduces the number of animals to be sacrificed and the statistical variability inherent in comparisons among different groups of animals. Animal models based on internal BL reporters enable researchers to perform in vivo efficacy, pharmacokinetics, toxicology, and target validation studies on many different classes of compounds. BL imaging challenges are currently moving towards two directions: multimodality imaging and the use of red and near infrared emitting probes to improve light penetration through tissues [67].

Several works compared the performance of different reporter genes in BL imaging, driving to different conclusions. For example, Luc2 was considered the best luciferase for in vivo neuroimaging, while a firefly luciferase mutant Ppy RE9 (GenBank accession number GQ404466) developed by Branchini was shown to be optimal for in vivo imaging of stem cells [68, 69].

In particular, the energy transfer from an enzyme to an inorganic nanomaterial led to revolutionary "lighting strategies." Rao et al. pioneered this approach, demonstrating the feasibility of conjugating semiconductive quantum dots (QDs) to mutated *R. reniformis* luciferase (Luc8) via EDC coupling. In the presence of the substrate coelenterazine, a non-radiative energy transfer (BRET) occurs. BRET allows conversion of the luciferase blue-green emission to the red and near-infrared emission typical of the quantum dot [70]. A typical limitation of this approach was the limited efficiency of the BRET phenomenon. Alam et al. proposed a BRET between a firefly variant (PpyGR-TS) and quantum dots to create bionanoconjugates [71] with potential applications in the fields of infrared signaling, sensing, and in vivo imaging.

By integrating BRET and FRET, Xiong et al. [72], demonstrated the suitability of near-infrared imaging without external light excitation for lymphonode mapping and cancer imaging in living mice, allowing one to image very small tumors of 2–3 mm in diameter.

6 Outlook

Luciferases have been widely used as efficient reporters to develop different types of assays, including cell-based assays, protein–protein, and protein–ligand interaction assays, in vitro single cell imaging, and in vivo assays. Such formidable proteins have been genetically engineered to emit at different wavelengths or with higher efficiency, allowing the development of highly sensitive and/or multiplexed assays. In addition, BL enables the analysis of very small volume samples, thus being suitable for miniaturized, high-throughput BL-based assays [74, 75].

The continuous advancements in the fields of BL let us envisage further BL exploitation, such as the discovery of new BL reporter genes and the production of luciferase mutants with improved spectral properties. Further advancements of BL applications will be also made possible by new instrumentation for light measurement, such as ultrasensitive color CCD cameras for the simultaneous detection of different light-emitting probes. Moreover, the enormous advances in nanotechnology and nanofabrication provided exciting "nanotools" for advanced diagnostics and theranostics relying on BL imaging, with the potential for opening completely new approaches.

References

1. Wilson T, Hastings JW (1998) Bioluminescence. Annu Rev Cell Dev Biol 14:197–230
2. Ando Y, Niwa K, Yamada N et al (2008) Firefly bioluminescence quantum yield and colour change by pH-sensitive green emission. Nat Photonics 2:44–47
3. Kricka LJ (2000) Application of bioluminescence and chemiluminescence in biomedical sciences. Methods Enzymol 305:333–345
4. Cali JJ, Niles A, Valley MP et al (2008) Bioluminescent assays for ADMET. Expert Opin Drug Metab Toxicol 4:103–120
5. Roda A, Michelini E, Cevenini L et al (2014) Integrating biochemiluminescence detection on smartphones: mobile chemistry platform for point-of-need analysis. Anal Chem. 86:7299–7304
6. Scott D, Dikici E, Ensor M, Daunert S (2011) Bioluminescence and its impact on bioanalysis. Annu Rev Anal Chem (Palo Alto Calif) 4:297–319
7. De Wet JR, Wood KV, Helinski DR et al (1985) Cloning of firefly luciferase cDNA and the expression of active luciferase in Escherichia coli. Proc Natl Acad Sci USA 82:7870–7873
8. Branchini BR, Southworth TL, DeAngelis JP et al (2006) Luciferase from the Italian firefly *Luciola italica*: molecular cloning and expression. Comp Biochem Physiol B Biochem Mol Biol 145:159–167

9. Markova SV, Golz S, Frank LA et al (2004) Cloning and expression of cDNA for a luciferase from the marine copepod Metridia longa. A novel secreted bioluminescent reporter enzyme. J Biol Chem 279:3212–3217
10. Stolz U, Velez S, Wood KV et al (2003) Darwinian natural selection for orange bioluminescent color in a Jamaican click beetle. Proc Natl Acad Sci USA 100:14955–14959
11. Li J, Chen L, Du L et al (2013) Cage the firefly luciferin!—a strategy for developing bioluminescent probes. Chem Soc Rev 42:662–676
12. Navizet I, Liu YJ, Ferré N et al (2011) The chemistry of bioluminescence: an analysis of chemical functionalities. Chemphyschem 12:3064–3076
13. Fraga H, Fontes R, Esteves da Silva JC (2005) Synthesis of luciferyl coenzyme A: a bioluminescent substrate for firefly luciferase in the presence of AMP. Angew Chem Int Ed Engl 44:3427–3429
14. Day RN, Kawecki M, Berry D (1998) Dual-function reporter protein for analysis of gene expression in living cells. Biotechniques 25:848–850, 852–854, 856
15. Leclerc GM, Boockfor FR, Faught WJ et al (2000) Development of a destabilized firefly luciferase enzyme for measurement of gene expression. Biotechniques 29(590-591):594–596
16. Dranchak P, MacArthur R, Guha R et al (2013) Profile of the GSK published protein kinase inhibitor set across ATP-dependent and-independent luciferases: implications for reporter-gene assays. PLoS One 8:e57888
17. Herbst KJ, Allen MD, Zhang J (2009) The cAMP-dependent protein kinase inhibitor H-89 attenuates the bioluminescence signal produced by Renilla Luciferase. PLoS One 4:e5642
18. Ho PI, Yue K, Pandey P, Breault L et al (2013) Reporter enzyme inhibitor study to aid assembly of orthogonal reporter gene assays. ACS Chem Biol 8:1009–1017
19. Branchini BR, Southworth TL, Khattak NF et al (2005) Red- and green-emitting firefly luciferase mutants for bioluminescent reporter applications. Anal Biochem 345:140–148
20. Kitayama A, Yoshizaki H, Ohmiya Y et al (2003) Creation of a thermostable firefly luciferase with pH-insensitive luminescent color. Photochem Photobiol 77:333–338
21. Dunlap P (2014) Biochemistry and genetics of bacterial bioluminescence. Adv Biochem Eng Biotechnol 144:37–64
22. Tannous BA, Teng J (2011) Secreted blood reporters: insights and applications. Biotechnol Adv 29:997–1003
23. Lewandrowski GK, Magee CN, Mounayar M et al (2014) Simultaneous in vivo monitoring of regulatory and effector T lymphocytes using secreted gaussia luciferase, Firefly luciferase, and secreted alkaline phosphatase. Methods Mol Biol 1098:211–227
24. Yamada Y, Nishide SY, Nakajima Y et al (2013) Monitoring circadian time in rat plasma using a secreted Cypridina luciferase reporter. Anal Biochem 439:80–87
25. Hall MP, Unch J, Binkowski BF et al (2012) Engineered luciferase reporter from a deep sea shrimp utilizing a novel imidazopyrazinone substrate. ACS Chem Biol 7:1848–1857
26. Chen Y, Wang L, Cheng X et al (2014) An ultrasensitive system for measuring the USPs and OTULIN activity using Nanoluc as a reporter. Biochem Biophys Res Commun 455:178–183
27. Heise K, Oppermann H, Meixensberger J et al (2013) Dual luciferase assay for secreted luciferases based on Gaussia and NanoLuc. Assay Drug Dev Technol 11:244–252
28. Stacer AC, Nyati S, Moudgil P et al (2013) NanoLuc reporter for dual luciferase imaging in living animals. Mol Imaging 12:1–13
29. Loh JM, Proft T (2014) Comparison of firefly luciferase and NanoLuc luciferase for biophotonic labeling of group A Streptococcus. Biotechnol Lett 36:829–834
30. Nakajima Y, Kobayashi K, Yamagishi K et al (2004) cDNA cloning and characterization of a secreted luciferase from the luminous Japanese ostracod, Cypridina noctiluca. Biosci Biotechnol Biochem 68:565–570
31. Markova SV, Larionova MD, Burakova LP et al (2014) The smallest natural high-active luciferase: Cloning and characterization of novel 16.5-kDa luciferase from copepod Metridia longa. Biochem Biophys Res Commun. doi:10.1016/j.bbrc.2014.12.082 (in press)
32. Kelkar M, De A (2012) Bioluminescence based in vivo screening technologies. Curr Opin Pharmacol 12:592–600

33. Nakajima Y, Ohmiya Y (2010) Bioluminescence assays: multicolor luciferase assay, secreted luciferase assay and imaging luciferase assay. Expert Opin Drug Discov 5:835–849
34. Michelini E, Cevenini L, Mezzanotte L et al (2009) Luminescent probes and visualization of bioluminescence. Methods Mol Biol 574:1–13
35. Baumann B, van der Meer JR (2007) Analysis of bioavailable arsenic in rice with whole cell living bioreporter bacteria. J Agric Food Chem 55:2115–2120
36. Ivask A, François M, Kahru A et al (2004) Recombinant luminescent bacterial sensors for the measurement of bioavailability of cadmium and lead in soils polluted by metal smelters. Chemosphere 55:147–156
37. Raut N, O'Connor G, Pasini P et al (2012) Engineered cells as biosensing systems in biomedical analysis. Anal Bioanal Chem 402:3147–3159
38. Rodriguez R, Miller KM (2014) Unravelling the genomic targets of small molecules using high-throughput sequencing. Nat Rev Genet 15:783–796
39. Kesarwala AH, Prior JL, Sun J et al (2006) Second-generation triple reporter for bioluminescence, micro-positron emission tomography, and fluorescence imaging. Mol Imaging 5:465–474
40. Date A, Pasini P, Sangal A, Daunert S (2010) Packaging sensing cells in spores for long-term preservation of sensors: a tool for biomedical and environmental analysis. Anal Chem 82:6098–6103
41. Cevenini L, Michelini E, D'Elia M et al (2013) Dual-color bioluminescent bioreporter for forensic analysis: evidence of androgenic and anti-androgenic activity of illicit drugs. Anal Bioanal Chem 405:1035–1045
42. Ekström L, Cevenini L, Michelini E et al (2013) Testosterone challenge and androgen receptor activity in relation to UGT2B17 genotypes. Eur J Clin Invest 43:248–255
43. Gammon ST, Leevy WM, Gross S et al (2006) Spectral unmixing of multicolored bioluminescence emitted from heterogeneous biological sources. Anal Chem 78:1520–1527
44. Cevenini L, Camarda G, Michelini E et al (2014) Multicolor bioluminescence boosts malaria research: quantitative dual-color assay and single-cell imaging in *Plasmodium falciparum* parasites. Anal Chem 86:8814–8821
45. Mezzanotte L, An N, Mol IM et al (2014) A new multicolor bioluminescence imaging platform to investigate NF-κB activity and apoptosis in human breast cancer cells. PLoS One. 9:e85550
46. Van Rijn S, Würdinger T, Nilsson J (2014) Multiplex functional bioluminescent reporters using Gaussia luciferase fused to epitope tags in an immunobinding assay. Methods Mol Biol 1098:231–247
47. Bolton EK, Sayler GS, Nivens DE et al (2002) Integrated CMOS photodetectors and signal processing for very low-level chemical sensing with the bioluminescent bioreporter integrated circuit. Sens Actuators B Chem 85:179–185
48. Jouanneau S, Durand MJ, Thouand G (2012) Online detection of metals in environmental samples: comparing two concepts of bioluminescent bacterial biosensors. Environ Sci Technol 46:11979–11987
49. Michelini E, Cevenini L, Calabretta MM et al (2013) Field-deployable whole-cell bioluminescent biosensors: so near and yet so far. Anal Bioanal Chem 405:6155–6163
50. Simpson ML, Sayler GS, Patterson G et al (2001) An integrated CMOS microluminometer for low-level luminescence sensing in the bioluminescent bioreporter integrated circuit. Sens Actuators B Chem 72:134–140
51. Welsh DK, Noguchi T (2012) Cellular bioluminescence imaging. Cold Spring Harb Protoc 8:2012
52. Yasunaga M, Nakajima Y, Ohmiya Y (2014) Dual-color bioluminescence imaging assay using green- and red-emitting beetle luciferases at subcellular resolution. Anal Bioanal Chem 406:5735–5742
53. De A, Jasani A, Arora R, Gambhir SS (2013) Evolution of BRET biosensors from live cell to tissue-scale in vivo imaging. Front Endocrinol (Lausanne) 4:131

54. Kaczor AA, Makarska-Bialokoz M, Selent J et al (2014) Application of BRET for studying G protein-coupled receptors. Mini Rev Med Chem 14:411–425
55. Lohse MJ, Nuber S, Hoffmann C (2012) Fluorescence/bioluminescence resonance energy transfer techniques to study G-protein-coupled receptor activation and signaling. Pharmacol Rev 64:299–336
56. Dacres H, Michie M, Anderson A et al (2013) Advantages of substituting bioluminescence for fluorescence in a resonance energy transfer-based periplasmic binding protein biosensor. Biosens Bioelectron 41:459–464
57. Shekhawat SS, Ghosh I (2011) Split-protein systems: beyond binary protein-protein interactions. Curr Opin Chem Biol 15:789–797
58. Takakura H, Hattori M, Tanaka M et al (2015) Cell-based assays and animal models for GPCR drug screening. Methods Mol Biol 1272:257–270
59. Paulmurugan R, Tamrazi A, Massoud TF et al (2011) In vitro and in vivo molecular imaging of estrogen receptor α and β homo- and heterodimerization: exploration of new modes of receptor regulation. Mol Endocrinol 25:2029–2040
60. Fan-Minogue H, Cao Z, Paulmurugan R et al (2010) Noninvasive molecular imaging of c-Myc activation in living mice. Proc Natl Acad Sci USA 107:15892–15897. doi:10.1073/pnas.1007443107
61. Koterba KL, Rowan BG (2006) Measuring ligand-dependent and ligand-independent interactions between nuclear receptors and associated proteins using bioluminescence resonance energy transfer (BRET). Nucl Recept Signal 4:e021
62. Hsu CY, Chen CW, Yu HP et al (2013) Bioluminescence resonance energy transfer using luciferase-immobilized quantum dots for self-illuminated photodynamic therapy. Biomaterials 34:1204–1212
63. Klerk CP, Overmeer RM, Niers TM et al (2007) Validity of bioluminescence measurements for noninvasive in vivo imaging of tumor load in small animals. Biotechniques 43(7–13):30
64. Sato A, Klaunberg B, Tolwani R (2004) In vivo bioluminescence imaging. Comp Med. 54:631–634
65. Tseng JC, Kung AL (2015) Quantitative bioluminescence imaging of mouse tumor models. Cold Spring Harb Protoc. doi:10.1101/pdb.prot078261
66. Michelini E, Cevenini L, Calabretta MM et al (2014) Exploiting in vitro and in vivo bioluminescence for the implementation of the three Rs principle (replacement, reduction, and refinement) in drug discovery. Anal Bioanal Chem 406:5531–5539
67. Badr CE (2014) Bioluminescence imaging: basics and practical limitations. Methods Mol Biol 1098:1–18
68. Liang Y, Walczak P, Bulte JW (2012) Comparison of red-shifted firefly luciferase Ppy RE9 and conventional Luc2 as bioluminescence imaging reporter genes for in vivo imaging of stem cells. J Biomed Opt 17:016004
69. Mezzanotte L, Aswendt M, Tennstaedt A et al (2013) Evaluating reporter genes of different luciferases for optimized in vivo bioluminescence imaging of transplanted neural stem cells in the brain. Contrast Media Mol Imaging 8:505–513
70. So MK, Xu C, Loening AM et al (2006) Self-illuminating quantum dot conjugates for in vivo imaging. Nat Biotechnol 24:339–343
71. Alam R, Karam LM, Doane TL (2014) Near infrared bioluminescence resonance energy transfer from firefly luciferase-quantum dot bionanoconjugates. Nanotechnology 25:495606
72. Xiong L, Shuhendler AJ, Rao J (2012) Self-luminescing BRET-FRET near-infrared dots for in vivo lymph-node mapping and tumour imaging. Nat Commun 3:1193
73. Michelini E, Cevenini L, Mezzanotte L et al (2010) Cell-based assays: fuelling drug discovery. Anal Bioanal Chem 398:227–238
74. Roda A, Cevenini L, Borg S et al (2013) Bioengineered bioluminescent magnetotactic bacteria as a powerful tool for chip-based whole-cell biosensors. Lab Chip 13:4881–4889
75. Roda A, Roda B, Cevenini L et al (2011) Analytical strategies for improving the robustness and reproducibility of bioluminescent microbial bioreporters. Anal Bioanal Chem 401:201–211

Adv Biochem Eng Biotechnol (2016) 154: 19–46
DOI 10.1007/10_2015_324

Published Online: 16 May 2015

Measurement of Bacterial Bioluminescence Intensity and Spectrum: Current Physical Techniques and Principles

Kun Jia and Rodica Elena Ionescu

Abstract Bioluminescence is light production by living organisms, which can be observed in numerous marine creatures and some terrestrial invertebrates. More specifically, bacterial bioluminescence is the "cold light" produced and emitted by bacterial cells, including both wild-type luminescent and genetically engineered bacteria. Because of the lively interplay of synthetic biology, microbiology, toxicology, and biophysics, different configurations of whole-cell biosensors based on bacterial bioluminescence have been designed and are widely used in different fields, such as ecotoxicology, food toxicity, and environmental pollution. This chapter first discusses the background of the bioluminescence phenomenon in terms of optical spectrum. Platforms for bacterial bioluminescence detection using various techniques are then introduced, such as a photomultiplier tube, charge-coupled device (CCD) camera, micro-electro-mechanical systems (MEMS), and complementary metal-oxide-semiconductor (CMOS) based integrated circuit. Furthermore, some typical biochemical methods to optimize the analytical performances of bacterial bioluminescent biosensors/assays are reviewed, followed by a presentation of author's recent work concerning the improved sensitivity of a bioluminescent assay for pesticides. Finally, bacterial bioluminescence as implemented in eukaryotic cells, bioluminescent imaging, and cancer cell therapies is discussed.

Keywords Bioluminescence · Genetically engineered bacteria · Luminometer · Pesticide toxicity · Whole-cell biosensor

K. Jia · R.E. Ionescu (✉)
Laboratoire de Nanotechnologie et d'Instrumentation Optique,
Institut Charles Delaunay, Université de Technologie de Troyes, UMR CNRS 6281,
12 rue Marie Curie CS 42060, TROYES 10004 Cedex, France
e-mail: elena_rodica.ionescu@utt.fr

Contents

1 Introduction

Bioluminescent bacterial sensor-reporters, which use genetically engineered microorganisms to produce a light signal proportional to toxicant content, have offered an interesting simple alternative for monitoring the environmental pollution since the 1990s [1]. Because of the diversity of sensing parts (promoter gene) [2] and reporting elements (fluorescent protein or luciferase) [3], as well as the intrinsic properties of living cells, whole-cell bioassays using engineered bacterial cells are able to detect the effects of bioavailable parts of toxicants and also quantify the total toxicity of several toxicant mixtures to living cells. The possible evaluation of bioavailability via these bacterial bioreporters is therefore regarded as the most prominent advantage over other traditional biosensing techniques using biomolecules (e.g. antibodies, enzymes, nucleotides) as sensing elements [4]. Moreover, bacterial bioluminescence, enabled by the *lux* genes, is one special kind of chemiluminescence that occurs in living microbial cells without the requirement of external optical excitation; thus, the absence of background luminescent signals from host cells indicates that bacteria bioluminescent detection efficiency is solely determined by the optical detector. Consequently, a bioluminescent assay can achieve extremely high sensitivity due to changes in the self-illumination of bioluminescence in the presence of any physical or chemical stress. Therefore, genetically modified microorganisms have been widely used in the screening of environmental toxicants, such as organic pollutants [5, 6], heavy metals [7–9], pesticides [10–13], and antibiotics [14, 15].

Often, recombinant bacterial cells have been constructed by transforming the plasmids fused with promoter probes (such as *recA* or *grpE*) and reporter genes (*luxCDABE*, *luxAB* or *luc*) into host cells (e.g. *Escherichia coli*). The promoter probe is responsible for the specific recognition of the analyte, while the expression of the reporter gene leads to the production of bioluminescence via the synthesis of luciferase and its corresponding substrate [16]. Depending on the toxicants to be detected, a large number of engineered bacteria strains harboring different promoter genes have been constructed, as summarized in review articles [17–21]. The most frequently used bioreporter *lux* genes responsible for encoding the luciferase enzyme are the *luxCDABE* gene [22–24]. For example, *E. coli* TV1061 (*grpE*::*luxCDABE*) is sensitive to cellular metabolism changes, which is attributed to promoter *grpE*, one of about 20 promoters that is activated under specific conditions when the cells experience protein damage (e.g. upon addition of ethanol or phenol). Following this activation, the reporter gene is transcribed into the necessary components that are responsible for light production [25]. Another frequently used bacterial strain is based on the fusion of a DNA-damage inducible promoter *recA* and reporter gene (*luxCDABE*) and is responsive in the presence of genotoxical materials [26].

The first part of this chapter presents some background information on bioluminescence in terms of the optical spectrum, followed by the introduction of various physical platforms where the bacterial bioluminescent signal is detected and analyzed. Moreover, several protocols to improve the analytical performance of bacterial bioluminescent bioassays or biosensors is discussed based on the literature.

The temperature for monitoring bacterial bioluminescence evolution is usually 25 °C. Based on this limitation, the authors investigated different temperatures and reported their results through two independent protocols that drastically improved the sensitivity of the bacterial bioluminescent biodetection of pesticides. Finally, some perspectives and emerging applications of bacterial bioluminescence beyond biochemical compound detection and toxicity are briefly discussed.

2 Bioluminescence Background

2.1 Intensity and Spectrum of Bioluminescence

Bioluminescence occurs due to the presence of a specific enzyme named luciferase. Biochemically, all known luciferases are oxygenases that oxidize their corresponding substrates (generically named luciferin, and literally meaning “light-bearing” molecules) in the presence of oxygen. The spectra properties of this bioluminescent emission are dependent on various parameters, such as luciferase/luciferin structures, organism habitat, optical biological filters, and accessory lumipores [e.g. green fluorescent protein (GFP) and yellow fluorescent protein (YFP)]. Taking GFP, in the bioluminescent marine hydroid *Obelia* as an example, it has been reported that light emissions from the living organism are green, whereas blue light is emitted during in vitro reaction. Such results indicate that the different emitted colors can be attributed to the different enzyme conformations under in vivo and in vitro conditions [27]. The color of the bioluminescent signal is strongly dependent on an organism's habitat. Thus, deep-sea species have blue emissions (450–490 nm), the bioluminescence of coastal marine species is green (490–520 nm), whereas terrestrial and fresh water species are red-shifted to 550–580 nm [28].

Although bioluminescence has been observed in various organisms, bacterial bioluminescence occurs naturally in 11 bacterial species from four genera (*Vibrio, Photobacterium, Shewanella, and Photorhabdus*). In terms of bioluminescent spectra, the majority of bacteria cells emit blue-green light with a single peak located around 490 nm, which is driven by a complex biochemical reaction involving a luciferase enzyme, aliphatic aldehyde, reduced flavin mononucleotide, and oxygen. However, experimentally, two bioluminescent peaks have been observed using a sensitive photodetector [29]. Thus, the first peak appeared at 490 nm when detected by a conventional spectrophotometer based on photomultipliers, whereas the second peak was visualized in the yellow range of 585–595 nm when using a more sensitive Raman CCD camera (Fig. 1). More interestingly, the emission spectra from different types of bacteria (nature vs. recombinant cells, Gram-positive vs. Gram-negative strains) shared similar properties; these emission spectra were also found to be independent of bacterial cell growth media and conditions. Thouand et al. [29] claimed that the second peak in the yellow range could be attributed to the presence of an accessory emitter protein inside bacterial cells and/or the autofluorescence of the luciferase enzyme.

The concurrent presence of two emission peaks at 490 and 540 nm from a single bacterial strain has been previously reported [30]. The YFP, which carries a flavin moiety, was found to be responsible for the appearance of a second peak and was temperature sensitive (+4 °C) both in vivo and in vitro, with changed in the color of bioluminescent emissions and a reaction rate that increased up to tenfold. Thus, the concomitant modification of emission spectra and reaction kinetics indicated that the emitter YFP should interact with *Lux* intermediate to form the excited state, which finally produces the second peak located in yellow range [31].

2.2 Instrumental Optics for Bacterial Bioluminescence

There are several methodologies and commercial instruments available for measurement of bacterial bioluminescence, generically named *luminometers*. A luminometer is used to record the optical signal from a bioluminescent assay in liquid

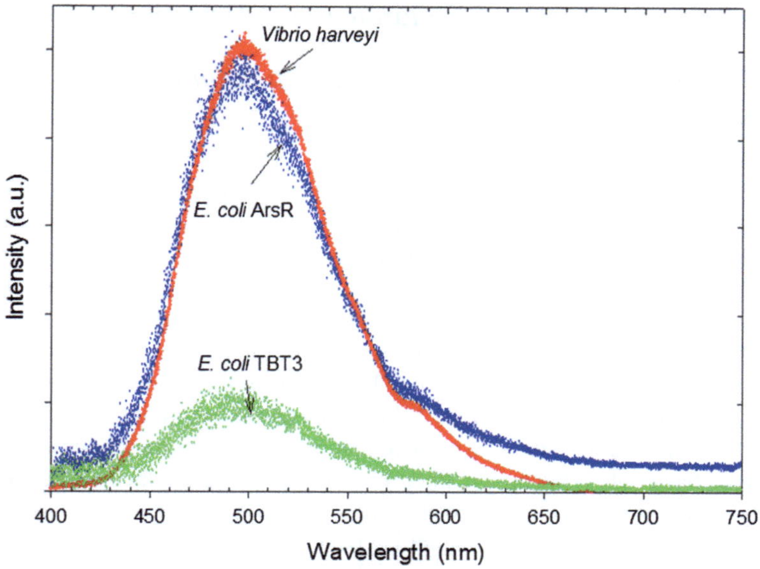

Fig. 1 Emission spectra of two engineered microorganisms strains (*E. coli* and *Vibrio harveyi*) cultivated in batch conditions in DSMZ6904 medium. Besides the normal peak at 500 nm, an additional peak around 585 nm was recorded for both strains (*E. coli* ArsR and *Vibrio harveyi*) using a more sensitive Raman CCD camera

phase. An imaging device is normally required if the optical signal is produced from two-dimensional sources, such as membranes used in blotting and tissue sections [32]. Specifically, a luminometer is designed for the measurement of a very small amount of light in the visible range. The photon detector frequently used is either a photomultiplier (PMT) or solid-state photodiode, where the detected signal may be either current or photon counts.

Different models of luminometers are available, either as a single-sample fitted with a PMT or solid-state detector or as multi-sample and automatic luminometers dedicated for microtiter plate measurement [33]. On the other hand, the imaging device permitting the location and measurement of light from a two-dimensional source is usually constructed by using a CCD camera as the detector and specialized software for imaging signal processing. There are also some hybrid instruments, such as a camera luminometer [33], where the samples in the microtiter plates are exposed to high-speed film. Semi-quantitative results can be obtained very quickly from this inexpensive instrument.

The following sections briefly discuss the various physical platforms that are available for bacterial bioluminescent detection.

3 Platforms for Bacterial Bioluminescent Detection

3.1 *Photomultiplier Tube for Bacterial Bioluminescent Detection*

One typical example of a vacuum phototube is the PMT tube, which is an extremely sensitive detector for light in a wide spectrum, ranging from ultraviolet to near-infrared wavelengths. Bacterial bioluminescence, enabled by the *lux* reporter, normally appears as a blue-green light with a center wavelength at 490 nm; thus, it can be easily detected by the PMT tube. For instance, optical fibers have been developed based on bioluminescent bioreporters for online water pollutant detection and air toxicity monitoring [34–37]. An illustration of the homemade setup for air toxicity monitoring is shown in Fig. 2. In this biosensor setup, the sensing element is a self-contained unit, with a disposable feature concerning the immobilization of engineered bacterial on the optical fiber tip cores via a calcium mediated gelation of alginate. The signal detection element is represented by a PMT tube. Upon exposure to a given toxicant, a light signal is created and transmitted through an optic fiber to the selected PMT. Thus, the analytical performances of the above system are optimized by investigating different experimental parameters: immobilized bacterial cells density, numerical aperture of optical fiber core, working temperature, etc., to detect 25 μg/L of mitomycin C [37]. Unlike this online biosensor configuration, others researchers [38] have immobilized the bacterial cells onto a nondisposable optical fiber-optical. For experiments, the genetically engineered *E. coli* bacteria cells containing firefly luciferase gene fused to *TOL* plasmid have been immobilized on optical fibers for benzene derivative detection at the ppm level with a Hamamatsu C2310 PMT [38].

A multiplexing bacterial luminescent bioassay using a bundle of 22 optical fibers was simultaneously prepared to detect various heavy metal pollutants [39]. Interestingly, it is reported that the disadvantages result from the cross-reaction among the presence of different metals. Furthermore, a portable optical biosensor based on immobilized bacterial cells, namely Lumisens2, was constructed by fixing the bacterial cells in a disposable card. The resulting bioluminescent signal was transmitted through an optical fiber to a PMT that enabled the detection of 2 μM for organotin compounds (tributyltin) in 400 min [40].

Although the bacterial bioluminescent signal can be readily detected by a conventional PMT with high sensitivity [37], the high working voltage, cost, large size, and the limited linear response range hinder its application in compact integrated biosensing systems. As a promising alternative, the solid-state silicon photomultiplier (SPM) shows some advantages over conventional PMT, such as small size, low bias voltage, and insensitivity to magnetic interference. Additionally, the SPM is more compatible with modern standard microelectronics technology containing a complementary metal oxide semiconductor (CMOS), which offers better integration of the optical signal detection element (SPM) and signal postprocessing unit on board [41, 42]. Followed by the initial work using the SPM to detect bacterial

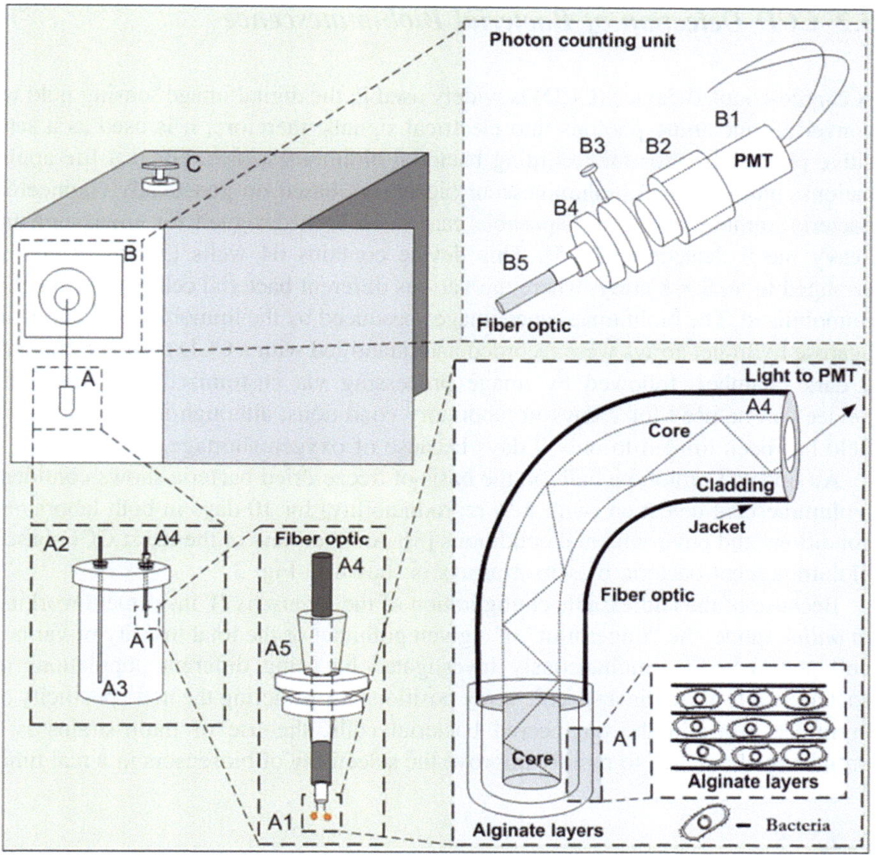

Fig. 2 Fiber-optic setup for the real-time monitoring of toxicity in the air. *A* detection unit; *A1* hermetic chamber; *A2* alginate matrix with bacteria immobilized on the fiber-optic core; *A3* needle for air pollutant entrance; *A4* Fiber optic; *A5* fiber-optic holder; *B* photon counting unit; *B1* Hamamatsu HC135-01 PMT sensor module; *B2* PMT fixation ring; *B3* manual shutter (71430, Oriel); *B4* fiber holder that prevents the movement of the fiber inside the photon counting unit; *B5* fiber optic; *C* outside handle of manual shutter that enables light access to the PMT. Reprinted with permission from [35], copyright 2011 Elsevier

bioluminescent signals [43], a more detailed study involving SPM detection of low-level bioluminescent signals was reported [44]. Based on these results, the authors claimed that the photo-counting readout properties and signal-to-noise ratio of SPM can be enhanced by using a thermoelectric cooler and digital filter, respectively. For model toxicant (salicylate) detection, the lower limit of detection of 250 ppb was obtained by using a digital filter to improve the signal-to-noise ratio. However, the implication of SPM or PMT in a fully integrated applicable biochip platform is still questionable, as the power consumption, size, and additional electronics need to be optimized.

3.2 CCD Detection of Bacterial Bioluminescence

A charge-coupled device (CCD) is widely used in the digital image sensing field by converting incoming photons into electrical signals; therefore, it is used as a sensitive photon detector for recording bacterial bioluminescence. For real-life applications, multi-channel bioluminescent biosensors based on genetically engineered bacteria immobilized in a disposable card have been designed for environmental heavy metal detection [8, 45]. This device contains 64 wells (3 mm in depth) arranged in an 8 × 8 array, where the various different bacterial cells can be locally immobilized. The bioluminescence images produced by the immobilized bacteria in agarose hydrogel arrays were recorded and quantified with a CCD cooled camera in a dark chamber, followed by image processing via customized software. This device can be used for 7 days in laboratory conditions, although its application in field has been limited to only 2 days because of oxygen shortage.

An updated prototype built on the basis of freeze-dried bacteria shows confident bioluminescent detection (with 3 % reproducibility) for 10 days in both laboratory conditions and environmental conditions [8]. An overview of the latest CCD-based bioluminescent bacteria biosensor arrays is shown in Fig. 3.

Because of the addressable configuration of the *Lumisens IV* instrument working in *online* mode, the "fingerprint" of a given pollutant or the total toxicity of various pollutants can be simultaneously investigated by using different populations of bacterial cells placed in a specific array positions. Considering the non-specificity of promoter genes in the engineered bacterial cells, the use of multi-strains is a promising alternative to rapidly improve the selectivity of biosensors in a real time.

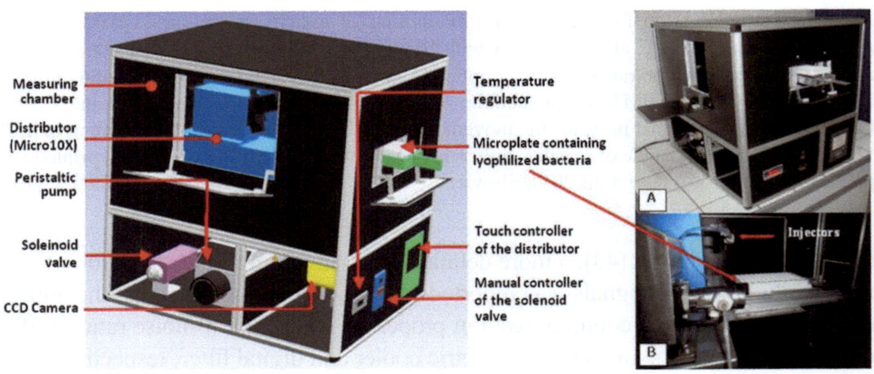

Fig. 3 Three-dimensional modeling performed by computer-aided design (*left*) and photos (*right*) of the Lumisens IV instrument (**a** global view; **b** inside of the measuring chamber). Reprinted with permission from Jouanneau et al. [8], copyright 2012 ACS

3.3 Electrochemical Detection of Bacterial Bioluminescence

A microfluidic whole-cell bacterial biosensor has been developed based on the electrochemical detection of water pollutants [46]. The biochip was fabricated by MEMS technology on a silicon substrate that contains four microchambers of different volumes and electrode radii (see Fig. 4). When engineered bacterial cells are exposed to the toxicant, a specific enzyme is produced, which catalyzes the enzymatic reaction of an exogenous substrate, leading to the generation of electrochemical active materials. Consequently, the concentration of toxicants can be directly correlated with the quantity of electroactive species created over cellular biochemical reactions. The authors assumed that a biotic-micro electro mechanical (biotic-MEMS) system would allow the fast screening of unknown analytes in a high-throughput manner.

Based on the above application of micro-electrochemical biochips in bacterial bioluminescent assays, the optimization of analytical performances has been reported [47]. Specifically, the authors developed two independent protocols for the modification of a working electrode integrated in an electrochemical biochip. Thus, one protocol used Cu–Au micropillars to increase the electrode's effective area, while the second one used conductive polypyrrole (PPy) films. It was noticed that,

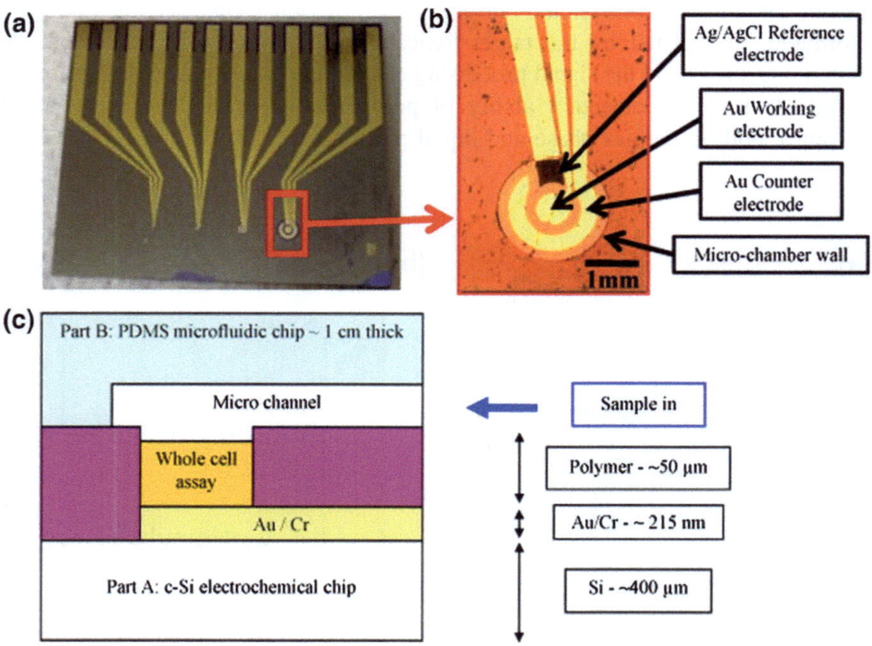

Fig. 4 A silicon-based micro-chip with four differentially sized electrochemical micro-chambers (**a**). Internal view of a single three-electrode electrochemical micro-chamber (**b**). Schematic layout of the electrochemical unit (**c**). Reprinted with permission from Ben-Yoav et al. [46], copyright 2009 Elsevier

in the presence of bioluminescent bacteria, both protocols produced a higher current signal when compared to the nonmodified working electrodes. Concerning the efficiency of the biodetection, the Cu/Au micropillar structured electrode provided a 24 % increase compared to the nonmodified electrode. However, the PPy modification of the working electrode resulted in an overall decrease of efficiency, which could be attributed to a high background signal.

3.4 CMOS Detection of Bacterial Bioluminescence

As discussed previously, although bacterial bioluminescence can be sensitively collected by different photon detectors, it is still a big challenge to design a compact and field-portable bacteria biosensor that includes all requested elements, such as incubation, detection, and signal processing. Despite these constraints, some research groups have designed field-deployable bioluminescent whole-cell biosensors based on the advanced CMOS technique [48, 49]. A highly compact integrated circuit CMOS micro-luminometer of low power (<100 μW), small size (1.5 × 1.5 mm), and on-board signal processing was reported [50, 51]. This micro-luminometer was fabricated by the standard 0.35-μm CMOS process, containing integrated photodiodes for bioluminescent detection and a signal processing unit to convert the current from photodiode arrays to a digital signal bearing frequency information modulated by the target toxicant concentrations (Fig. 5). Both the photodiode arrays and the signal processing electronic circuits have been optimized to provide sensitive detection ability of pollutants (salicylate and naphthalene, 0.1 ppm), both in the gas phase and liquid phase.

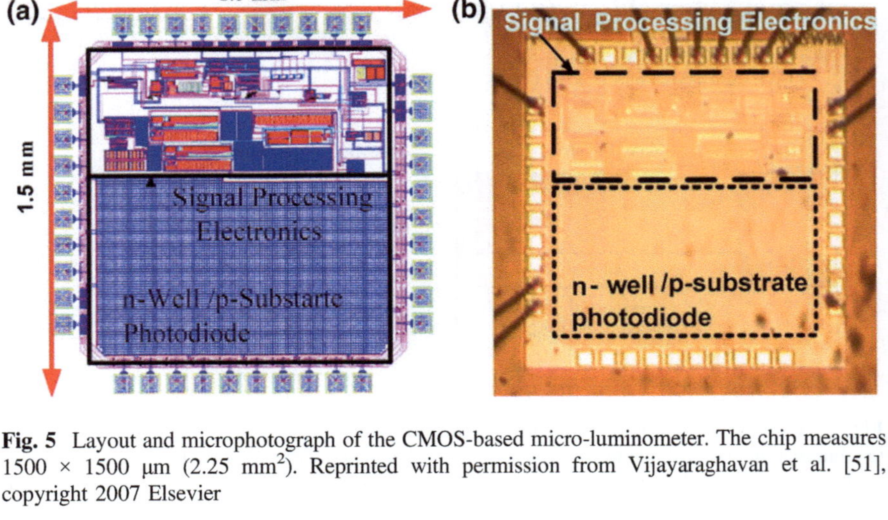

Fig. 5 Layout and microphotograph of the CMOS-based micro-luminometer. The chip measures 1500 × 1500 μm (2.25 mm^2). Reprinted with permission from Vijayaraghavan et al. [51], copyright 2007 Elsevier

3.5 Bacterial Bioreporter Arrays for High-Throughput Measurements

An important aspect in the bioluminescent field is the development of appropriate protocols for the immobilization of living bacterial cells onto different solid transducers, (optical fibers, electrochemical biochips, lab-on-chip total microanalysis), which will be of potential interest for designing high-throughput and sensitive detection of analyte(s). Therefore, a wide range [17, 19, 20] of bacteria array biosensors have been created by assembling a panel of bacterial cells harboring different promoter:reporter fusions. For instance, bacteria sensor arrays were fabricated by printing various engineered bacteria cell spots onto glass surfaces using an adapted noncontact robotic printer [52]. Experimentally, the bacterial cell immobilization efficiency can be improved either by chemical modification of the substrate surface with amino group functionalized self-assembled monolayers or biological modification of bacteria cells resulting in an overproduction of external cellular surface protein by gene mutation. It has been reported that the viability and bioluminescent activity of the immobilized bacterial cells can be preserved for at least 2 months when kept at 4 °C. The optical microscope images of bacteria arrays on different glass supports are shown in Fig. 6.

Generally, bacterial bioluminescence is derived from the luciferase-catalyzed bioreaction involving three substrates: oxygen, reduced riboflavin 5'-phosphate ($FMNH_2$), and long-chain aliphatic aldehyde. It is known that the emission spectrum of natural bacterial bioluminescence has a wide peak, ranging from 400 to 650 nm with a maximum absorption at 490 nm. However, bacterial bioluminescent color can present fluctuations. For instance, the bacterial bioluminescent peak can be red-shifted to 530 nm or blue-shifted to 470 nm by replacing the original $FMNH_2$ substrate with 2-thiol FMN and iso-FMN, respectively [53]. Another method to affect the bacterial spectrum is by using random mutation [54] or site-directed mutagenesis [55]. However, the bacterial activity is substantially reduced and only a small shift of bioluminescent spectrum is obtained by the above-mentioned methods. Thus, under in vivo and in vitro conditions, bacteria bioluminescence can be affected by the formation of a non-covalent complex between luciferase and specific fluorescent proteins (e.g. lumazine protein from *Photobacterium phosphoreum* [56], yellow- and blue-fluorescent proteins from *Alivibrio fischeri* strain Y1 [57, 58]. Due to the limited formation of luciferase-bacterial fluorescent protein complex [59] and its non-covalent nature, the bioluminescent spectrum is very sensitive to external perturbation factors, such as temperature and concentration. Besides these procedures, one robust and versatile protocol for investigating the bioluminescent resonant energy transfer (BRET) effect between *Vibro harveyi* luciferase (energy donor) and coral *Discosoma sp.* fluorescent protein mOrange (energy receptor) has been reported [60], where an

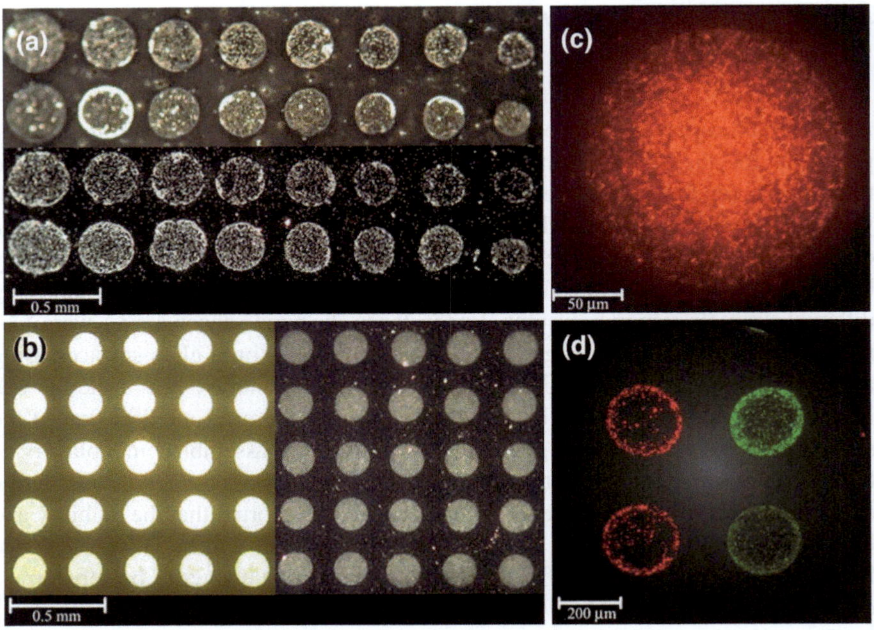

Fig. 6 Light microscope image of fluorescent reporter strains SM110 and SM111 printed (0.5–2.5 nL) on a non-modified glass slide. Printing (*upper* 2 rows) and 24 h incubation at 30 °C were followed by the removal of non-adhered cells (2 *bottom* rows) (**a**). Strain SM111 (1 nL) was printed on APTES-coated glass (5 *left* columns) and incubated for 24 h at 4 °C, followed by the removal of non-adhered cells (5 *right* columns) (**b**). Epifluorescence microscope images of strains SM110 (**c**) and SM111 (**d**) were printed (2 nL) on APTES-coated glass. Reprinted with permission from [52], copyright 2011 RSC

obvious modification of the bacterial bioluminescent spectrum could be observed (Fig. 7). This protocol has some advantages: the fluorescent protein is not directly exposed to the laser excitation that is usually used in conventional fluorescence and the emission from the bacterial luciferase-fluorescent protein fusion is derived from an enzymatic reaction [61].

In summary, due to a specific bioluminescent response signal in the presence of a particular toxicant, the bioreporter whole-cell chips are easy-to-handle tools for identification of a wide spectrum of toxicants [62] and compounds from unknown mixtures [63] that help in the classification of their toxicity [64, 65]. However, only a few studies have investigated the change of bioluminescent light color from wild-type luciferase systems for future applications in the development of multiplexing whole-cell biosensors [66].

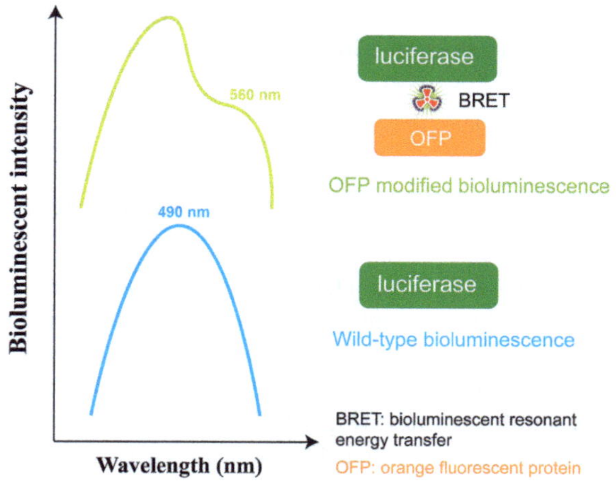

Fig. 7 The bioluminescence spectra from wild-type luciferase (from *Vibrio harveyi*) is displayed as the *bottom blue curve* containing a single peak at 490 nm, while an additional shoulder peak at 560 nm is detected for the luciferase covalently modified by orange fluorescent protein (OFP, from coral *Discosoma sp.*), shown as the *top green curve*

4 Protocols for Improving The Analytical Performances of Bacterial Bioluminescence

4.1 Protocol Based on The Optimization of Bacterial Culture Medium

Since the early stages of bioluminescence investigations, the optimization of bacterial cell medium has been considered as a principal parameter responsible for high signal-response events. Thus, by comparing the toxicant-induced bioluminescent signals from bacterial cells grown in four mediums containing different carbon sources (acetate, pyruvate, glucose, and citrate), Thouand's group found that bacteria cultivated in a medium with acetate exhibited the highest sensitivity, which was attributed to the alternation of bacteria's metal exchange and adsorption via changing energy sources and bacterial cellular membrane states [67]. The induction ratios of bacterial bioluminescence are also dependent on the growth phase and the genetic construction of bacterial cells, as different promoters are expressed in varying degrees under different growth phases.

Even though the promoter gene of most genetically modified bacteria is responsive to a range of heavy metals, the selectivity of the whole-cell bioluminescent assays could be improved by exposing various bacterial cells to the same samples at the same time. Thereby, the combination of different bacterial responses will decrease the detection time of targets. In other cases, the bacterial cells need immobilization and the choice of substrate is an important issue. An ideal "bioluminescent" substrate should have strong adhesion ability for the bacterial cells, have no chemical interaction with the target compound(s), and be transparent and biocompatible.

4.2 Protocol Based on Electrochemical Measurements

Since most bacterial cells are surface charged, the external electric field can also be used to enhance the bioluminescent response in the whole-cell biosensors using electrochemical platforms. For instance, a portable bacteria whole-cell biochip was fabricated by controlling the adhesion of genetically modified bacterial cells onto the ITO-based electrochemical substrate (Fig. 8). The exposure of bacterial cells to target toxicants initiates a cascade of cellular biochemical reactions, leading to the luciferase enzyme that catalyzes a specific added substrate with the production of electrochemical active species detected by a chrono-amperometric technique [68, 69].

4.3 Protocol Based on Electrophoretic Effects

An interesting procedure for the immobilization of bacterial cells on conductive ITO surfaces using the electrophoretic deposition (EPD) technique has recently been reported [70, 71]. The EPD technique consists of depositing charged particles

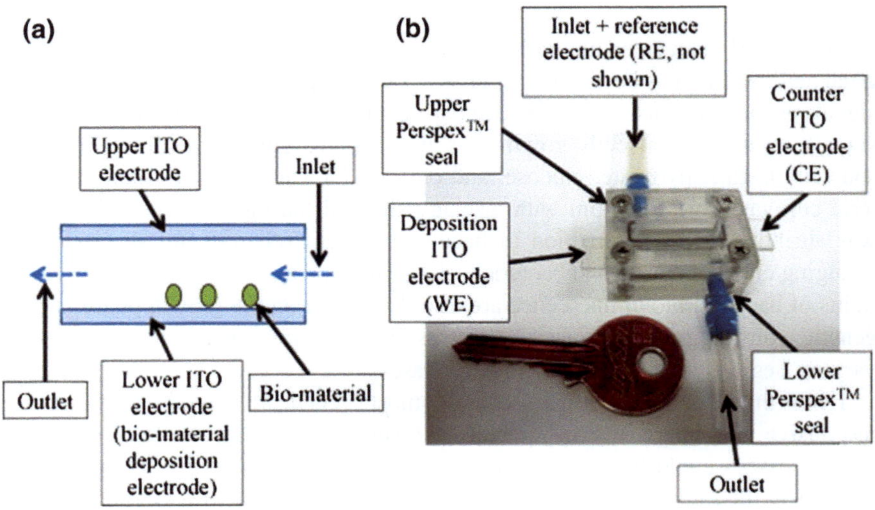

Fig. 8 Electrochemical-based flow chamber designed for microscopic observations and electrochemical measurements with the flow chamber *side view* (**a**) and the entire assembled flow chamber (**b**). Reprinted with permission from Ben-Yoav et al. [69], copyright 2011 Elsevier

in the liquid phase onto a solid conductive surface with an opposite charge under a DC or AC electric field. For cells, the adhesion phenomenon is attributed to the random attachment of cells on supports followed by nucleation and growth of cell cluster biological events. Thus, the efficiency of electrophoretic bio-deposition (EPBD) is highly dependent on the zeta potential and the electrophoretic mobility of bacterial cells. Practically, even though the electrostatic forces between negatively charged bacterial cells and positively charged ITO electrodes exist, the electrophoretic force established between two parallel electrodes will be the driving force for manipulation of bacterial cells [68].

Because the sensitivity of a biosensor is mainly determined by the diffusion rate of analytes inside the bacterial cells [72], the electroporation approach will greatly contribute to the deformation of bacterial membranes, thus increasing permeability to external molecules [73, 74].

4.4 Protocol Based on Molecular Manipulation

The essential ingredient of genetically engineered bioluminescent bacteria is the fusion of bacterial bioluminescent genes (*lux*) and promoter genes. The latter is responsible for the specific recognition of target toxicants or stress effects exposed to bacterial cells, whereas the expression of the former produces dose-dependent light emission. The promoter gene is a non-coding DNA sequence preceding the actual gene coding section, while the transcription of downstream genes is only initiated after this promoter gene is recognized by the RNA polymerase. Therefore, the appropriate molecular manipulation of promoter region should improve the total analytical performance of bacterial bioluminescent sensors.

An interesting approach that used four independent strategies to manipulate the promoter gene has been reported by Yagur-Kroll et al. [75]. Specifically, the first strategy was to modify the length of the DNA segment containing the promoter region. The second approach introduced a random gene mutation via a directed evolution process, whereas the third approach introduced more specific site mutations in the promoter sequence (−35 and −10 regions). The final strategy was the duplication of the promoter sequence to increase the binding sites for RNA polymerase. As a consequence, the bioluminescent sensitivity, kinetic response (earlier response time), and intensity of emitted light were significantly improved. Belkin's group claimed that these molecular manipulation protocols were applicable to almost all of the promoter:reporter fusions of genetically engineered bacterial cells.

In a typical promoter:reporter fusion, the emitted optical signal is derived from the expression of the *lux* gene, which is regulated by the upstream promoter sequence. Thus, besides molecular manipulation of promoter region to improve bioluminescent efficiency, another genetic method has been reported to update the reported bacterial bioluminescence performance by splitting the *lux* gene [76]. Specifically, the *luxCDABE* cassette from *Photorhabdus luminescens* was divided into two smaller subunits of *luxAB* and *luxCDE*, with the former encoding the enzyme luciferase and the latter encoding an enzymatic complex responsible for synthesizing the long-chain aldehyde substrate of the bioluminescent reaction. These two smaller subunits can be genetically modified to be under control of an inducible stress-responsive promoter or a synthetic constitutive promoter. The authors tested the different combinations of inductive/constitutive *luxAB* and inductive/constitutive *luxCDE*; they found that the optimized configuration was an inducible *luxAB* and a constitutive *luxCDE*, which indicated that the stronger intensity and faster response of bacterial bioluminescent reporter in this case may be attributed to the improved bioavailability of the aliphatic aldehyde substrate. The other possible reason is that the transcription and/or translation of short DNA segments may be faster and more efficient than that of long ones after lux operon splitting. This study clearly proved the previously published results that aliphatic aldehyde substrate is the rate-limiting factor of bioluminescent reactions when the entire *luxCDABE* gene is used [77].

In addition to the molecular manipulation of the *lux* bioreporter and its promoter sequence, other methodologies based on molecular biology have been developed. For instance, a faster bioluminescent response was recorded either by using different plasmid-bearing host strains [78]. Also, the background signal of an *arsR*::*lacZ* bioreporter was greatly reduced by placing another *ArsR* binding site downstream of *arsR* [79].

To conclude, the bioluminescent sensitivity can be increased by certain methods, such as integration of the promoter:reporter fusion into the bacterial chromosome instead of plasmid, modifying the cell permeability towards the target compound, and decreasing the plasmid copy number [80, 81].

4.5 Protocol Based on Numerical Modelling

The specificity of a bacterial bioluminescent assay is dictated by the biorecognition element of a promoter sequence inside the cells, although this response circuit can be tightly controlled to only respond to a single compound [79, 82]. It is known that most bioluminescent bacteria bioreporters are normally responsive to a broad spectrum of chemicals sharing similar structures [83]. To improve the analytical specificity of a bacterial bioluminescent assay, a simple alternative is to use a panel of several bacteria reporters containing the *luxCDABE* operon fused with different stress promoters. When various bacteria strains are exposed to target compounds, "fingerprint" bioluminescent responses of each target are produced by the corresponding bioreporter, which are analyzed by a special computing program based on different algorithms. Some examples of algorithms for identification of antibiotics and heavy metals are the Bayesian decision theory, nonparametric nearest-neighbor technique [84, 85], chi-squared automatic interaction detector type [7], and support vector machine classifier [86].

4.6 Protocol Based on Overnight "Cold Incubation"

To improve the sensitivity of a bioluminescent assay for the detection of water pollutants (e.g. atrazine), a new protocol was created using a two-stage bioluminescence reading approach (both were required): one for continuous monitoring of light evolution for 5 h (fresh stage investigation) and a second for another 5 h (after an "overnight" cold incubation at 4 °C) [11]. In both phases, the bioluminescence responses were acquired at 25 °C. The "overnight" cold incubation condition was tested for different periods of times, such as 3, 6, 9, and 12 h, respectively. From the obtained results, 9 h at 4 °C gave the best sensitivity for atrazine bioluminescent detection.

Due to the importance of the bacterial growth phase, intact bacterial cells of different optical density (OD) were first bioluminescent tested at 25 °C (Fig. 9). Interestingly, it was found that for the fresh mode (right-side group), two peaks were always recorded, no matter what OD was selected (0.03, 0.08 or 0.12). Moreover, when the OD increased from 0.03 to 0.12, a significant shift of the peaks from 265 to 210 min was observed. The increase of the OD has also led to an increase of the peak amplitude. By using the overnight mode (left-side group), larger bioluminescent peaks were recorded after 30 min, where the appearances of such peaks were OD independent. However, the intensity of the "overnight" peak

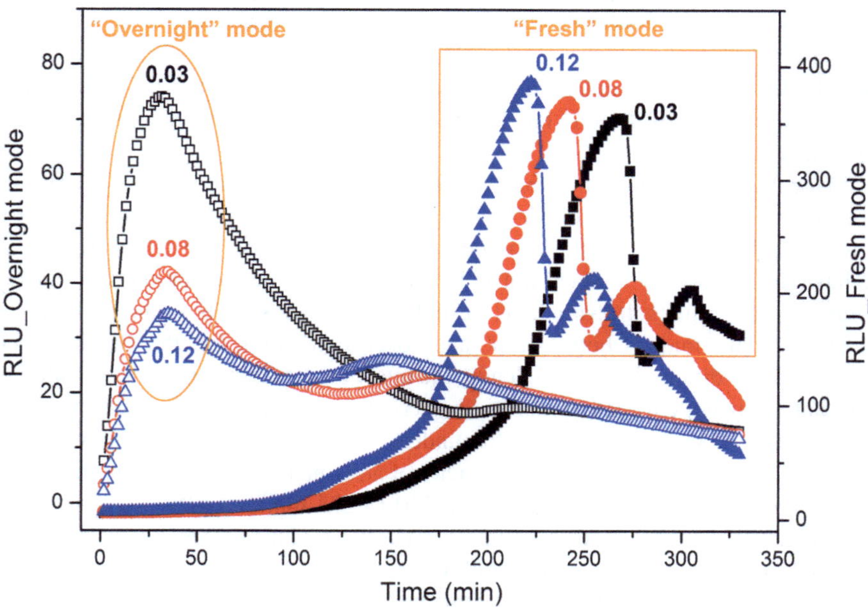

Fig. 9 Bioluminescence of *E. coli* TV1061 bacteria of different cell optical densities (0.03: *black*, 0.08: *red* and 0.12: *blue*) within "fresh" and "overnight" conditions. Both types of bacterial bioluminescence were recorded at 25 °C. An optimized cold incubation (4 °C) was obtained after 9 h

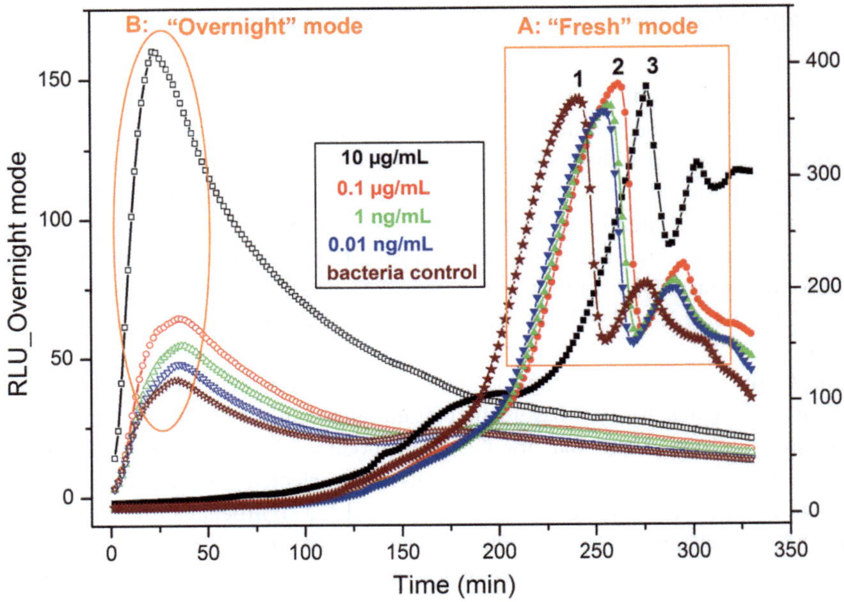

Fig. 10 Bioluminescence behaviors of *E. coli* TV1061 bacteria (OD = 0.08) within the presence of different atrazine concentrations recorded in fresh mode for 5.5 h (group A) and after an overnight incubation of *E. coli* TV1061 at 4 °C (group B). Four atrazine concentrations (10 µg/mL, 0.1 µg/mL, 1 ng/mL and 0.01 ng/mL) were tested for their toxicity to bacterial cells. Bioluminescent measurements for the two modes were recorded at 25 °C

was decreased when compared to the major peak obtained within the fresh mode. Thus, the increase of the OD has led to a decrease of the peak amplitude for the overnight mode.

Once the optimization of the bacterial OD value was acquired, the toxic effect of atrazine compound to *E. coli* TV1061 bacteria was investigated through the proposed "fresh" and "overnight" consecutive tests. Bacteria of 0.08 OD were proven to have the best ability to discriminate between different concentrations of atrazine. Thus, Fig. 10 depicts the variations in bioluminescence signals in the presence of atrazine versus time for both "fresh" and "overnight" tests. For the "fresh" mode (group A), conditions 1, 2, and 3 denote the intact bacteria, bacteria treated with low toxicant (atrazine) concentrations, and bacteria treated with high toxicant (atrazine) concentrations, respectively. Spectra under conditions 1 and 2 exhibited very similar shapes. For condition 3, the toxicity of atrazine is bioluminescent, translated as an important peak shift in time (from 240 to 275 min).

For the "overnight" mode (group B), bacterial bioluminescence was rationally dependent on the atrazine concentration. In contrast to the overlapped bacterial response to lower concentrations of atrazine obtained in "fresh mode," larger distinguished peaks were obtained after only 25 min for various low concentrations of atrazine (from 10 fg/mL to 10 µg/mL). Although it is difficult to elucidate the specific mechanism of bacterial bioluminescence evolution under the overnight conditions, the use of cold incubation supposedly slows down the bacteria's own metabolisms or may enable the accumulation of the toxicant inside the cells. Consequently, the bacteria are gradually chemical "stressed" by the increase of the atrazine content.

4.7 Protocol Based on "Washing" Bacterial Cells

The generation of bioluminescence is a dynamic process highly dependent on cellular metabolism status [87]. An easy-to-perform protocol has been reported by Jia et al. [10]. The protocol contains three major steps: incubation at 25 °C, centrifugation, and washing of aged cells with fresh LB medium, (Fig. 11). This affects

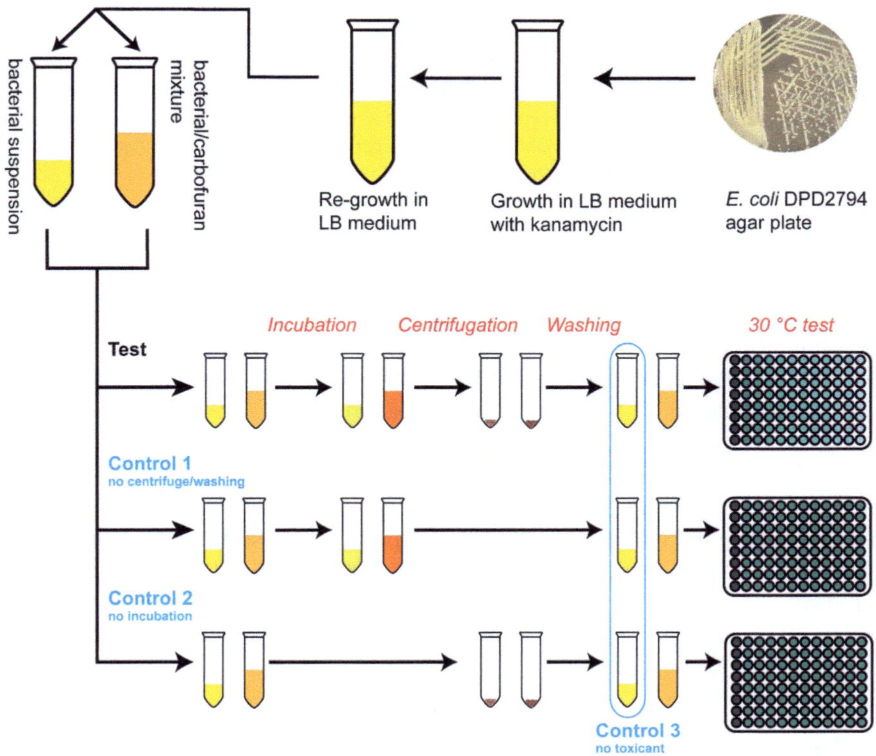

Fig. 11 A bioluminescence-enhancement protocol based on three steps: incubation of bacterial cells with a toxicant solution at room temperature for different periods of time, centrifugation of the resulted bacterial/toxicant suspension, and replacement of aged supernatant with a fresh LB medium free of toxicants. Finally, the bacterial bioluminescence was investigated at 30 °C. Three independent control experiments were prepared: first without the centrifugation/washing step, second without the incubation step, and third without any toxicants

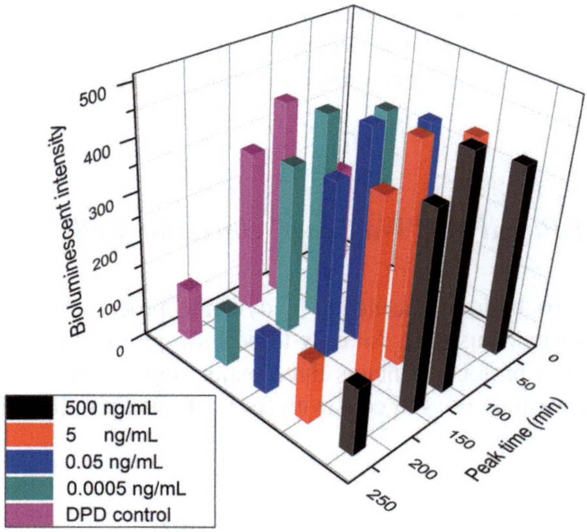

Fig. 12 Bioluminescent signals of DPD2794 bacterial cells induced by the presence of different carbofuran content (0.5 μg/mL, 5 ng/mL, 50 pg/mL, 0.5 pg/mL) and the signal of control cell samples (named DPD control). The incubation time of control bacteria and bacteria with toxicants was 6 h at room temperature, followed by independent centrifugation and washing steps with fresh LB medium before the final bioluminescent investigation at 30 °C

the *E. coli* DPD2794 cellular physiological response, for sensitive bioluminescent detection of a specific toxicant (carbofuran) at an optimized temperature (30 °C).

The precise incubation time of bacterial cells with toxicants and their subsequent washing with fresh LB medium after centrifugation were investigated as a function of the resulting bioluminescent signal in terms of its shape and intensity. Thus, from the five tested times of 0 h (no toxicant incubation), 2, 4, 6, 9, and 12 h at room temperature, 6 h proved to have the highest influence on the ability of *E. coli* DPD2794 cells to differentiate between different carbofuran concentrations (0.5 μg/mL, 5 ng/mL, 50 pg/mL, 0.5 pg/mL) after the bioluminescent test at 30 °C (Fig. 12).

Concerning the bioluminescent shape-response signal at two independent temperatures, a remarkable evolution was noted using the fresh-LB medium washing protocol, demonstrated by the appearance of broadened peaks after 250 min by using 30 °C and only after 350 min by using 26 °C [26, 88–90].

5 Emerging Bacterial Bioluminescence Applications Beyond Biosensing/Assays

Because of the extensive applications of whole-cell bacterial bioluminescence in biochemical and toxicity detection, the ongoing research in *lux*-based systems has been pushed from the original prokaryotic bacterial *luxCDABE* gene cassette further to eukaryotic cells. The first breakthrough of expressing *lux* genes in eukaryotic hosts used autonomous bioluminescence from *Saccharomyces cerevisiae* yeast [91], where more stable luciferase proteins from terrestrial bacterium *P. luminescens* were expressed instead of marine strains such as *Vibrio harveyi* or *Alivibrio fischeri*. Since this discovery, the expression of bacterial *lux* genes in eukaryotic cells has been under investigation. The development of a robust bioassay based on luminescent yeast, with the potential for *lux* gene expression in human cell lines [92, 93], has been reported.

In addition, the bioluminescent imaging (BLI) approach is becoming a powerful tool for the real-time and in vivo monitoring of various biological processes in living systems [61]. Thus, many bacterial cells can be genetically engineered with *lux* reporter genes to induce the production of light, dramatically enlarging the spectrum of in vivo bacterial investigations with this non-aggressive and easy-to-perform technique. Moreover, due to the fact that bacterial cells possess a natural ability to grow preferentially within tumor(s), engineered bioluminescent bacterial cells can be designed as theranostic probes for the identification of cancer cells. For instance, *Salmonella typhimurium* bacterial cells were genetically engineered to carry cytotoxic proteins (cytolysin A) and luminescent reporter (*lux*) genes for monitoring the cancer therapy process inside mice [94]. Fortunately, the resolution of in vivo bioluminescent imaging from tumor-targeted bacterial cells can be greatly enhanced by using a combinational three-dimensional diffuse optical tomography (3D BLI) and micro-computed tomography (μCT), as reported by Cronin et al. [95].

Besides cancer therapy and imaging investigations, bacterial bioluminescence can also be used in other scenarios, such as the study of intestinal colonization by *E. coli* in mice [96], as well as the monitoring and elucidation of antimicrobial photodynamic inactivation mechanisms [97, 98].

6 Conclusions and Perspectives

This chapter discussed several existing physical platforms used within bioluminescence measurements, such as PMT tubes, CCD cameras, electrochemical biochips, and CMOS-integrated circuits. Then, several published works were introduced on the improvement of the analytical performances of bacterial bioluminescence biosensing/assays. Moreover, the authors' recent work on the highly improved sensitivity of bacterial bioluminescent assays based on two protocols for the fine-tuning of bacterial physiology and application of *lux* genes in eukaryotes cells by Sayler's group were discussed as well.

Although bioluminescence bioassays for toxicant detection based on genetically engineered bacterial cells have been studied since the 1990s [1], much work remains to be done for their real-world application in different scenarios. The most challenging question is how to obtain applicable selectivity towards specific target pollutants, given the fact that the promoter gene used in most engineered bacteria normally is responsive to many toxicants. It is believed that the solution will be given by advancements in molecular biology, which will help us to understand the cellular interaction with exogenous compounds. Meanwhile, the fine tuning of bacterial metabolism by modulating various experimental parameters has already been proven as an economical option to enhance the analytical performance of bioassays.

Finally, the combination of a bioluminescent signal with other advanced devices, such as micro-/nano-fluidics, or advanced optical materials, such plasmonic gold nanoparticles, to further improve the efficiency of a whole test system will definitely push the current laboratorial bioluminescent research to real-time applications.

Acknowledgments The authors are grateful to the French Ministry of Foreign Affairs and the Israeli Ministry of Science for funding their project "Nano-chip platform for water probing pollutants" under the Research Networks Programs 2009–2011 and to the University of Technology of Troyes (UTT) for the Programme Stratégique 2009–2012. Kun Jia thanks the China Scholarship Council (CSC) for his PhD fellowship 2010–2013.

References

1. King J, DiGrazia P, Applegate B, Burlage R, Sanseverino J, Dunbar P, Larimer F, Ga Sayler (1990) Rapid, sensitive bioluminescent reporter technology for naphthalene exposure and biodegradation. Science 249(4970):778–781
2. Marqués S, Aranda-Olmedo I, Ramos JL (2006) Controlling bacterial physiology for optimal expression of gene reporter constructs. Curr Opin Biotechnol 17(1):50–56
3. Drepper T, Eggert T, Circolone F, Heck A, Krauß U, Guterl J-K, Wendorff M, Losi A, Gärtner W, Jaeger K-E (2007) Reporter proteins for in vivo fluorescence without oxygen. Nat Biotechnol 25(4):443–445
4. Magrisso S, Erel Y, Belkin S (2008) Microbial reporters of metal bioavailability. Microbial Biotechnol 1(4):320–330
5. Hay AG, Rice JF, Applegate BM, Bright NG, Sayler GS (2000) A bioluminescent whole-cell reporter for detection of 2, 4-dichlorophenoxyacetic acid and 2, 4-dichlorophenol in soil. Appl Environ Microbiol 66(10):4589–4594
6. Valdman E, Battaglini F, Leite S, Valdman B (2004) Naphthalene detection by a bioluminescence sensor applied to wastewater samples. Sens Actuat B-Chem 103(1):7–12
7. Jouanneau S, Durand M-J, Courcoux P, Blusseau T, Thouand G (2011) Improvement of the identification of four heavy metals in environmental samples by using predictive decision tree models coupled with a set of five bioluminescent bacteria. Environ Sci Technol 45 (7):2925–2931
8. Jouanneau S, Durand MJ, Thouand G (2012) Online detection of metals in environmental samples: comparing two concepts of bioluminescent bacterial biosensors. Environ Sci Technol 46(21):11979–11987
9. Liao VH-C, Chien M-T, Tseng Y-Y, Ou K-L (2006) Assessment of heavy metal bioavailability in contaminated sediments and soils using green fluorescent protein-based bacterial biosensors. Environ Pollut 142(1):17–23
10. Jia K, Eltzov E, Marks RS, Ionescu RE (2013) Bioluminescence enhancement through an added washing protocol enabling a greater sensitivity to carbofuran toxicity. Ecotoxicol Environ Saf 96:61–66
11. Jia K, Eltzov E, Toury T, Marks RS, Ionescu RE (2012) A lower limit of detection for atrazine was obtained using bioluminescent reporter bacteria via a lower incubation temperature. Ecotoxicol Environ Saf 84:221–226

12. Lee S, Suzuki M, Tamiya E, Karube I (1992) Sensitive bioluminescent detection of pesticides utilizing a membrane mutant of *Escherichia coli* and recombinant DNA technology. Anal Chim Acta 257(2):183–188
13. Pham CH, Min J, Gu MB (2004) Pesticide induced toxicity and stress response in bacterial cells. Bull Environ Contam Tox 72(2):380–386
14. Eltzov E, Ben-Yosef DZ, Kushmaro A, Marks R (2008) Detection of sub-inhibitory antibiotic concentrations via luminescent sensing bacteria and prediction of their mode of action. Sens Actuat B-Chem 129(2):685–692
15. Tenhami M, Hakkila K, Karp M (2001) Measurement of effects of antibiotics in bioluminescent *Staphylococcus aureus* RN4220. Antimicrob Agents Chemother 45 (12):3456–3461
16. Meighen EA (1991) Molecular biology of bacterial bioluminescence. Microbiol Rev 55 (1):123–142
17. Girotti S, Ferri EN, Fumo MG, Maiolini E (2008) Monitoring of environmental pollutants by bioluminescent bacteria. Anal Chim Acta 608(1):2–29
18. Lei Y, Chen W, Mulchandani A (2006) Microbial biosensors. Anal Chim Acta 568(1):200–210
19. Robbens J, Dardenne F, Devriese L, De Coen W, Blust R (2010) *Escherichia coli* as a bioreporter in ecotoxicology. Appl Microbiol Biotechnol 88(5):1007–1025
20. Su L, Jia W, Hou C, Lei Y (2011) Microbial biosensors: a review. Biosens Bioelectron 26 (5):1788–1799
21. Woutersen M, Belkin S, Brouwer B, van Wezel AP, Heringa MB (2011) Are luminescent bacteria suitable for online detection and monitoring of toxic compounds in drinking water and its sources? Anal Bioanal Chem 400(4):915–929
22. Abd-El-Haleem D, Ripp S, Scott C, Sayler G (2002) A luxCDABE-based bioluminescent bioreporter for the detection of phenol. J Ind Microbiol Biotechnol 29(5):233–237
23. Hwang ET, Ahn J-M, Kim BC, Gu MB (2008) Construction of a nrdA: luxCDABE fusion and its use in *Escherichia coli* as a DNA damage biosensor. Sensors 8(2):1297–1307
24. Lopes N, Hawkins SA, Jegier P, Menn F-M, Sayler GS, Ripp S (2012) Detection of dichloromethane with a bioluminescent (lux) bacterial bioreporter. J Ind Microbiol Biotechnol 39(1):45–53
25. Van Dyk TK, Majarian WR, Konstantinov KB, Young RM, Dhurjati PS, Larossa RA (1994) Rapid and sensitive pollutant detection by induction of heat shock gene-bioluminescence gene fusions. Appl Environ Microbiol 60(5):1414–1420
26. Vollmer AC, Belkin S, Smulski DR, Van Dyk TK, LaRossa RA (1997) Detection of DNA damage by use of *Escherichia coli* carrying recA': lux, uvrA': lux, or alkA': lux reporter plasmids. Appl Environ Microbiol 63(7):2566–2571
27. Morin JG, Hastings J (1971) Energy transfer in a bioluminescent system. J Cell Physiol 77 (3):313–318
28. Hastings J, Morin J (1991) Bioluminescence. In: Prosser CL (ed) Neural and integrative animal physiology. Wiley, New York, NY, pp 131–170
29. Thouand G, Daniel P, Horry H, Picart P, Durand MJ, Killham K, Knox OG, DuBow MS, Rousseau M (2003) Comparison of the spectral emission of lux recombinant and bioluminescent marine bacteria. Luminescence 18(3):145–155
30. Ruby EG, Nealson KH (1977) A luminous bacterium that emits yellow light. Science 196 (4288):432–434
31. Hastings J (1996) Chemistries and colors of bioluminescent reactions: a review. Gene 173 (1):5–11
32. Stanley PE (1997) Commercially available luminometers and imaging devices for low-light level measurements and kits and reagents utilizing bioluminescence or chemiluminescence: survey update 5. J Biolumin Chemilumin 12(2):61–78
33. Stanley PE (1992) A survey of more than 90 commercially available luminometers and imaging devices for low-light measurements of chemiluminescence and bioluminescence, including instruments for manual, automatic and specialized operation, for HPLC, LC, GLC and microtiter plates. Part1: descriptions. J Biolumin Chemilumin 7(2):77–108

34. Eltzov E, Marks RS, Voost S, Wullings BA, Heringa MB (2009) Flow-through real time bacterial biosensor for toxic compounds in water. Sens Actuat B-Chem 142(1):11–18
35. Eltzov E, Pavluchkov V, Burstin M, Marks RS (2011) Creation of a fiber optic based biosensor for air toxicity monitoring. Sens Actuat B-Chem 155(2):859–867
36. Polyak B, Bassis E, Novodvorets A, Belkin S, Marks R (2000) Optical fiber bioluminescent whole-cell microbial biosensors to genotoxicants. Water Sci Technol 42(1–2):305–311
37. Polyak B, Bassis E, Novodvorets A, Belkin S, Marks RS (2001) Bioluminescent whole cell optical fiber sensor to genotoxicants: system optimization. Sens Actuat B-Chem 74(1–3):18–26
38. Ikariyama Y, Nishiguchi S, Koyama T, Kobatake E, Aizawa M, Tsuda M, Nakazawa T (1997) Fiber-optic-based biomonitoring of benzene derivatives by recombinant *E. coli* bearing luciferase gene-fused TOL-plasmid immobilized on the fiber-optic end. Anal Chem 69 (13):2600–2605
39. Hakkila K, Green T, Leskinen P, Ivask A, Marks R, Virta M (2004) Detection of bioavailable heavy metals in EILATox-Oregon samples using whole-cell luminescent bacterial sensors in suspension or immobilized onto fibre-optic tips. J Appl Toxicol 24(5):333–342
40. Horry H, Charrier T, Durand M-J, Vrignaud B, Picart P, Daniel P, Thouand G (2007) Technological conception of an optical biosensor with a disposable card for use with bioluminescent bacteria. Sens Actuat B-Chem 122(2):527–534
41. Buzhan P, Dolgoshein B, Filatov L, Ilyin A, Kantzerov V, Kaplin V, Karakash A, Kayumov F, Klemin S, Popova E (2003) Silicon photomultiplier and its possible applications. Nucl Instrum Methods Phys Res Sect A 504(1):48–52
42. Stewart A, Saveliev V, Bellis S, Herbert D, Hughes P, Jackson J (2008) Performance of 1-mm^2 silicon photomultiplier. IEEE Quantum Electron 44(2):157–164
43. Daniel R, Almog R, Ron A, Belkin S, Diamand YS (2008) Modeling and measurement of a whole-cell bioluminescent biosensor based on a single photon avalanche diode. Biosens Bioelectron 24(4):882–887
44. Li H, Lopes N, Moser S, Sayler G, Ripp S (2012) Silicon photomultiplier (SPM) detection of low-level bioluminescence for the development of deployable whole-cell biosensors: possibilities and limitations. Biosens Bioelectron 33(1):299–303
45. Charrier T, Chapeau C, Bendria L, Picart P, Daniel P, Thouand G (2011) A multi-channel bioluminescent bacterial biosensor for the on-line detection of metals and toxicity. Part II: technical development and proof of concept of the biosensor. Anal Bioanal Chem 400 (4):1061–1070
46. Ben-Yoav H, Biran A, Pedahzur R, Belkin S, Buchinger S, Reifferscheid G, Shacham-Diamand Y (2009) A whole cell electrochemical biosensor for water genotoxicity bio-detection. Electrochim Acta 54(25):6113–6118
47. Ben-Yoav H, Ofek Almog R, Sverdlov Y, Sternheim M, Belkin S, Freeman A, Shacham-Diamand Y (2012) Modified working electrodes for electrochemical whole-cell microchips. Electrochim Acta 82:109–114
48. Bolton EK, Sayler GS, Nivens DE, Rochelle JM, Ripp S, Simpson ML (2002) Integrated CMOS photodetectors and signal processing for very low-level chemical sensing with the bioluminescent bioreporter integrated circuit. Sens Actuat B-Chem 85(1–2):179–185
49. Simpson ML, Sayler GS, Applegate BM, Ripp S, Nivens DE, Paulus MJ, Jellison GE Jr (1998) Bioluminescent-bioreporter integrated circuits form novel whole-cell biosensors. Trends Biotechnol 16(8):332–338
50. Islam SK, Vijayaraghavan R, Zhang M, Ripp S, Caylor SD, Weathers B, Moser S, Terry S, Blalock BJ, Sayler GS (2007) Integrated circuit biosensors using living whole-cell bioreporters. IEEE Trans Circuits Syst I Regul Pap 54(1):89–98
51. Vijayaraghavan R, Islam SK, Zhang M, Ripp S, Caylor S, Bull ND, Moser S, Terry SC, Blalock BJ, Sayler GS (2007) A bioreporter bioluminescent integrated circuit for very low-level chemical sensing in both gas and liquid environments. Sens Actuat B-Chem 123(2):922–928
52. Melamed S, Ceriotti L, Weigel W, Rossi F, Colpo P, Belkin S (2011) A printed nanolitre-scale bacterial sensor array. Lab Chip 11(1):139–146

53. Mitchell G, Hastings JW (1969) The effect of flavin isomers and analogues upon the color of bacterial bioluminescence. J Biol Chem 244(10):2572–2576
54. Cline TW, Hastings JW (1974) Mutated luciferases with altered bioluminescence emission spectra. J Biol Chem 249(14):4668–4669
55. Lin LY-C, Szittner R, Friedman R, Meighen EA (2004) Changes in the kinetics and emission spectrum on mutation of the chromophore-binding platform in *Vibrio harveyi* luciferase. Biochemistry 43(11):3183–3194
56. Lee J, O'Kane DJ, Gibson BG (1989) Bioluminescence spectral and fluorescence dynamics study of the interaction of lumazine protein with the intermediates of bacterial luciferase bioluminescence. Biochemistry 28(10):4263–4271
57. Karatani H, Hastings J (1993) Two active forms of the accessory yellow fluorescence protein of the luminous bacterium vibrio fischeri strain Y1. J Photochem Photobiol B: Biol 18 (2):227–232
58. Karatani H, Wilson T, Hastings J (1992) A blue fluorescent protein from a yellow-emitting luminous bacterium. Photochem Photobiol 55(2):293–299
59. Petushkov VN, Ketelaars M, Gibson BG, Lee J (1996) Interaction of photobacterium leiognathi and vibrio fischeri Y1 luciferases with fluorescent (antenna) proteins: bioluminecence effects of the aliphatic additive. Biochemistry 35(37):12086–12093
60. Ke D, Tu SC (2011) Activities, kinetics and emission spectra of bacterial luciferase-fluorescent protein fusion enzymes. Photochem Photobiol 87(6):1346–1353
61. Xu X, Soutto M, Xie Q, Servick S, Subramanian C, von Arnim AG, Johnson CH (2007) Imaging protein interactions with bioluminescence resonance energy transfer (BRET) in plant and mammalian cells and tissues. P Natl Acad Sci USA 104(24):10264–10269
62. Ahn J-M, Kim JH, Kim JH, Gu MB (2010) Randomly distributed arrays of optically coded functional microbeads for toxicity screening and monitoring. Lab Chip 10(20):2695–2701
63. Kim B, Gu M (2005) A multi-channel continuous water toxicity monitoring system: its evaluation and application to water discharged from a power plant. Environ Monit Assess 109 (1–3):123–133
64. Elad T, Almog R, Yagur-Kroll S, Levkov K, Melamed S, Shacham-Diamand Y, Belkin S (2011) Online monitoring of water toxicity by use of bioluminescent reporter bacterial biochips. Environ Sci Technol 45(19):8536–8544
65. Lee JH, Mitchell RJ, Kim BC, Cullen DC, Gu MB (2005) A cell array biosensor for environmental toxicity analysis. Biosens Bioelectron 21(3):500–507
66. Rowe L, Dikici E, Daunert S (2009) Engineering bioluminescent proteins: expanding their analytical potential. Anal Chem 81(21):8662–8668
67. Charrier T, Durand MJ, Jouanneau S, Dion M, Pernetti M, Poncelet D, Thouand G (2011) A multi-channel bioluminescent bacterial biosensor for the on-line detection of metals and toxicity. Part I: design and optimization of bioluminescent bacterial strains. Anal Bioanal Chem 400(4):1051–1060
68. Ben-Yoav H, Amzel T, Sternheim M, Belkin S, Rubin A, Shacham-Diamand Y, Freeman A (2011) Signal amelioration of electrophoretically deposited whole-cell biosensors using external electric fields. Electrochim Acta 56(26):9666–9672
69. Ben-Yoav H, Freeman A, Sternheim M, Shacham-Diamand Y (2011) An electrochemical impedance model for integrated bacterial biofilms. Electrochim Acta 56(23):7780–7786
70. Besra L, Liu M (2007) A review on fundamentals and applications of electrophoretic deposition (EPD). Prog Mater Sci 52(1):1–61
71. Neirinck B, Van Mellaert L, Fransaer J, Van der Biest O, Anné J, Vleugels J (2009) Electrophoretic deposition of bacterial cells. Electrochem Commun 11(9):1842–1845
72. Elad T, Lee JH, Gu MB, Belkin S (2010) Microbial cell arrays. Adv Biochem Eng Biotechnol 117:85–108
73. Ibey BL, Mixon DG, Payne JA, Bowman A, Sickendick K, Wilmink GJ, Roach WP, Pakhomov AG (2010) Plasma membrane permeabilization by trains of ultrashort electric pulses. Bioelectrochemistry 79(1):114–121

74. Weaver JC, Chizmadzhev YA (1996) Theory of electroporation: a review. Bioelectrochem Bioenerget 41(2):135–160
75. Yagur-Kroll S, Bilic B, Belkin S (2010) Strategies for enhancing bioluminescent bacterial sensor performance by promoter region manipulation. Microb Biotechnol 3(3):300–310
76. Yagur-Kroll S, Belkin S (2011) Upgrading bioluminescent bacterial bioreporter performance by splitting the lux operon. Anal Bioanal Chem 400(4):1071–1082
77. Fernández-Pinas F, Wolk CP (1994) Expression of luxCD-E in *Anabaena sp.* can replace the use of exogenous aldehyde for in vivo localization of transcription by luxAB. Gene 150 (1):169–174
78. Davidov Y, Rozen R, Smulski DR, Van Dyk TK, Vollmer AC, Elsemore DA, LaRossa RA, Belkin S (2000) Improved bacterial SOS promoter::lux fusions for genotoxicity detection. Mutat Res Genet Toxicol Environ Mutagenesis 466(1):97–107
79. Stocker J, Balluch D, Gsell M, Harms H, Feliciano J, Daunert S, Malik KA, van der Meer JR (2003) Development of a set of simple bacterial biosensors for quantitative and rapid measurements of arsenite and arsenate in potable water. Environ Sci Technol 37(20):4743–4750
80. Maehana K, Tani H, Shiba T, Kamidate T (2004) Effects of using a low-copy plasmid and controlling membrane permeability in SOS-based genotoxic bioassay. Anal Chim Acta 522 (2):189–195
81. Shapiro E, Baneyx F (2002) Stress-based identification and classification of antibacterial agents: second-generation Escherichia coli reporter strains and optimization of detection. Antimicrob Agents Chemother 46(8):2490–2497
82. Rasmussen LD, Sørensen SJ, Turner RR, Barkay T (2000) Application of a mer-lux biosensor for estimating bioavailable mercury in soil. Soil Biol Biochem 32(5):639–646
83. Ahn J-M, Gu M (2012) Geno-Tox: cell array biochip for genotoxicity monitoring and classification. Appl Biochem Biotechnol 168(4):752–760
84. Elad T, Benovich E, Magrisso S, Belkin S (2008) Toxicant identification by a luminescent bacterial bioreporter panel: application of pattern classification algorithms. Environ Sci Technol 42(22):8486–8491
85. Melamed S, Lalush C, Elad T, Yagur-Kroll S, Belkin S, Pedahzur R (2012) A bacterial reporter panel for the detection and classification of antibiotic substances. Microbial Biotechnol 5(4):536–548
86. Smolander O-P, Ribeiro AS, Yli-Harja O, Karp M (2009) Identification of β-lactam antibiotics using bioluminescent *Escherichia coli* and a support vector machine classifier algorithm. Sens Actuat B-Chem 141(2):604–609
87. Der Meer V, Roelof J, Tropel D, Jaspers M (2004) Illuminating the detection chain of bacterial bioreporters. Environ Microbiol 6(10):1005–1020
88. Belkin S, Smulski DR, Dadon S, Vollmer AC, Van Dyk TK, Larossa RA (1997) A panel of stress-responsive luminous bacteria for the detection of selected classes of toxicants. Water Res 31(12):3009–3016
89. Gu M, Choi S (2001) Monitoring and classification of toxicity using recombinant bioluminescent bacteria. Water Sci Technol 43(2):147–154
90. Premkumar JR, Rosen R, Belkin S, Lev O (2002) Sol–gel luminescence biosensors: encapsulation of recombinant *E. coli* reporters in thick silicate films. Anal Chim Acta 462 (1):11–23
91. Gupta RK, Patterson SS, Ripp S, Simpson ML, Sayler GS (2003) Expression of the *Photorhabdus luminescens* lux genes (luxA, B, C, D, and E) in *Saccharomyces cerevisiae*. FEMS Yeast Res 4(3):305–313
92. Close D, Xu T, Smartt A, Rogers A, Crossley R, Price S, Ripp S, Sayler G (2012) The evolution of the bacterial luciferase gene cassette (lux) as a real-time bioreporter. Sensors 12 (1):732–752
93. Close DM, Patterson SS, Ripp S, Baek SJ, Sanseverino J, Sayler GS (2010) Autonomous bioluminescent expression of the bacterial luciferase gene cassette (lux) in a mammalian cell line. PLoS One 5(8):e12441

94. Nguyen VH, Kim H-S, Ha J-M, Hong Y, Choy HE, Min J-J (2010) Genetically engineered *Salmonella typhimurium* as an imageable therapeutic probe for cancer. Cancer Res 70(1):18–23
95. Cronin M, Akin AR, Collins SA, Meganck J, Kim J-B, Baban CK, Joyce SA, van Dam GM, Zhang N, van Sinderen D (2012) High resolution in vivo bioluminescent imaging for the study of bacterial tumour targeting. PLoS One 7(1):e30940
96. Foucault M-L, Thomas L, Goussard S, Branchini B, Grillot-Courvalin C (2010) In vivo bioluminescence imaging for the study of intestinal colonization by *Escherichia coli* in mice. Appl Environ Microbiol 76(1):264–274
97. Alves E, Costa L, Cunha Â, Faustino MAF, Neves MGP, Almeida A (2011) Bioluminescence and its application in the monitoring of antimicrobial photodynamic therapy. Appl Microbiol Biotechnol 92(6):1115–1128
98. Tavares A, Dias SR, Carvalho CM, Faustino MA, Tomé JP, Neves MG, Tomé AC, Cavaleiro JA, Cunha Â, Gomes NC (2011) Mechanisms of photodynamic inactivation of a gram-negative recombinant bioluminescent bacterium by cationic porphyrins. Photochem Photobiolog Sci 10(10):1659–1669

Adv Biochem Eng Biotechnol (2016) 154: 47–74
DOI 10.1007/10_2014_281

Published Online: 7 December 2015

Structure, Mechanism, and Mutation of Bacterial Luciferase

Ruchanok Tinikul and Pimchai Chaiyen

Abstract Bacterial luciferase is a flavin-dependent monooxygenase found in bioluminescent bacteria. The enzyme catalyzes a light-emitting reaction by using reduced flavin, long chain aldehyde, and oxygen as substrates and yields oxidized flavin, carboxylic acid, and water as products with concomitant emission of blue-green light around 485–490 nm. The enzyme is a heterodimer consisting of two homologous subunits, designated as the α- and β-subunits. The reactive reaction center is located in the α-subunit, whereas the β-subunit is required for maintaining the active conformation of the α-subunit. The enzyme reaction occurs through the generation of a reactive C4a-oxygenflavin adduct, presumably C4a-peroxyflavin, before the light-emitting species is generated from the decomposition of an adduct between the C4a-peroxyflavin and the aldehyde. Because the luciferase reaction generates light, the enzyme has the potential to be used as a bioreporter for a wide variety of applications. With the recent invention of the fusion enzyme that can be expressed in mammalian cells, future possibilities for the development of additional bioreporter applications are promising.

Keywords Bacterial luciferase · Flavin-dependent monooxygenase · Enzyme mechanism · Enzyme kinetics · Mutagenesis

Contents

R. Tinikul
Mahidol University, Nakhonsawan Campus, Nakhonsawan 60130, Thailand

P. Chaiyen (✉)
Department of Biochemistry and Center of Excellence in Protein Structure and Function, Faculty of Science, Mahidol University, Bangkok 10400, Thailand
e-mail: pimchai.cha@mahidol.ac.th

1 Introduction

Bacterial luciferase catalyzes the oxidation of reduced flavin mononucleotide ($FMNH_2$, $FMNH^-$) and a long-chain aldehyde by molecular oxygen to yield oxidized FMN, a carboxylic acid and water with concomitant blue-green light ($\lambda_{max} \sim 490$ nm) emission. $FMNH_2 + RCHO + O_2 \rightarrow FMN + RCOOH + H_2O + h\nu\ (490\ nm)$ The bioluminescence quantum yield of bacterial luciferase is approximately 10–16 % [44, 60]. Most of these enzymes are found in Gram-negative bacteria [51]. The well-studied bacterial luciferases are from three genera, *Vibrio* and *Photobacterium* from marine environments [18, 21, 31] and *Photorhabdus* (*Xenorhabdus*) from terrestrial habitats [66].

Bacterial luciferase is encoded by adjacent *luxA* and *luxB* genes in the *lux* operon. In luminous bacteria, a long-chain aldehyde is synthesized by a fatty acid reductase multienzyme complex—the gene products of *luxC, luxD,* and *luxE*, which are encoded in the same *lux* operon. *luxD* encodes for a transferase that catalyzes the hydrolysis of a fatty acyl group to generate a free fatty acid. The gene products of *luxC* and *luxE* are a reductase and a synthase, respectively. They catalyze reduction of the fatty acid to an aldehyde, which serves as a substrate for the luciferase reaction. The NADH:FMN oxidoreductase encoded by the *luxG* gene in the *lux* operon is responsible for generating the majority of the $FMNH_2$ for the luminescence reaction [55] (Fig. 1). Tetradecanal (myristaldehyde) is believed to be the natural aldehyde substrate for luciferase because it is the aldehyde that is found in large amounts in the extract from luminous bacteria [60]. This notion is supported by the fact that mutants with deficiencies in the myristaldehyde biosynthesis pathway show a dim-light phenotype, and light emission can be stimulated by addition of tetradecanoic acid. Addition of tetradecanal or tetradecanoic acid in limited amounts together with respiration-blocking agents was found to increase light emission in the mutant by about 60-fold [71]. However, many long-chain aldehydes ranging from octanal to dodecanal can also be used as in vitro substrates by bacterial luciferase; however, different aldehydes yield different luminescence intensity [25].

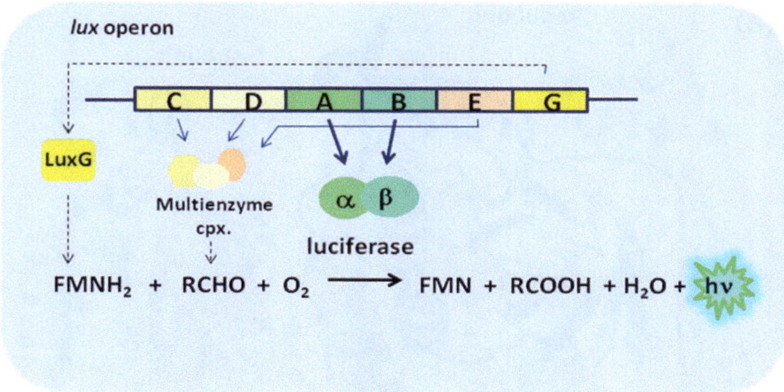

Fig. 1 The genes in the *lux* operon encode for various enzymes that function together to produce *blue-green* light in luminous bacteria

2 Structure of Bacterial Luciferase

Bacterial luciferase is a heterodimeric enzyme that is composed of asymmetric subunits designated as α- and β-subunits encoded by the *luxA* and *luxB* genes, respectively. The native molecular masses of enzymes from different species vary within a range of 77–80 kDa. The α-subunits (40–42 kDa) are generally larger than the β-subunits (36–37.5 kDa). The two types of subunits share about 30–32 % sequence identity amongst each other, with more than 95 of the 350 amino acids conserved [24, 62]. A high degree of sequence homology is also found among luciferases from different bacterial species (50–90 % identity). The sequence similarity is more conserved among luciferases with similar light decay kinetics (see Sect. 3.3). The enzymes from *V. harveyi, V. campbellii,* and *P. luminescens,* which are slow light decay enzymes, exhibit sequence identities as high as 84–90 %, while only about 60 % identity is shared among these enzymes and the enzymes from *Photobacterium* sp. (fast light decay) [18, 31, 66].

Both the α- and β-subunits of bacterial luciferase have TIM (β/α)8 barrel folding (Fig. 2a). A highly conserved sequence and structure among the α- and β-subunits suggest that their genes have a common origin. The enzyme active site is located solely in the α-subunit, while the β-subunit is required for the stabilization and maintenance of the active structure of the α-subunit [12]. When a homodimer of α_2 was generated, it was shown to have very low activity relative to the heterodimer [14]. The crystal structure of *V. harveyi* luciferase in complex with oxidized FMN indicates that the enzyme active site is situated at the α-subunit barrel (Fig. 2a). The *cis*-peptide bond between Ala74–Ala75 and the residues Cys106, Val173, and Ile191 (residues in magenta, Fig. 2b) are found at the *re*-face side of the isoalloxazine ring. The rare *cis*-peptide in the luciferase active site is very distinctive and has been proven to be important for the luciferase reaction [24, 45, 47]. The hydrophobic residues Trp194, Phe6, and Ser227 that form a hydrophobic surface

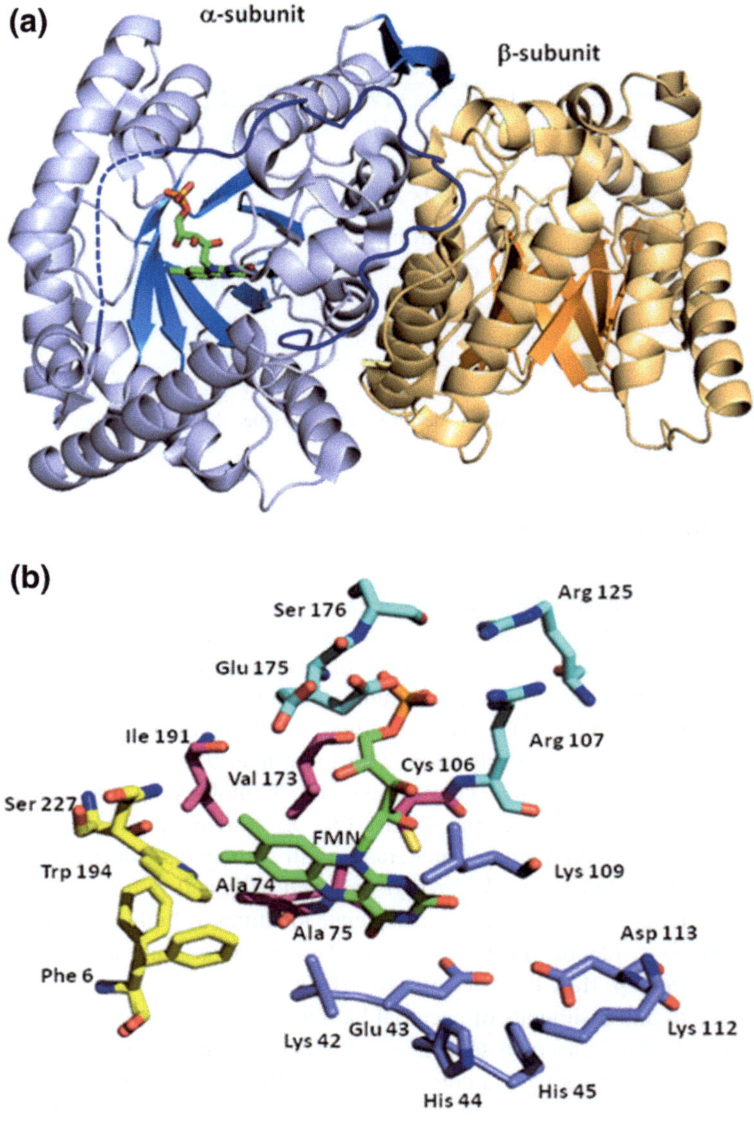

Fig. 2 **a** Structure of bacterial luciferase with FMN bound (pdb: 3FGC). The mobile loop segment is indicated in *dark blue*. **b** Residues around the FMN binding site. Phe6 and Glu175 exist in two conformations

are found at the *si*-face side of the isoalloxazine moiety near the dimethyl benzenoid ring. This pocket was proposed to be an aldehyde binding site (residues in yellow, Fig. 2b).

The pyrimidine moiety near the β-subunit interface is lined with polar residues (residues in blue, Fig. 2b), which form hydrogen bond networks. The electrostatic properties of this side of the cavity appear to be crucial for enzymatic function. The phosphate group of FMN is surrounded by the side chains of Arg107, Arg125, Glu175, and Ser 176 (residues in cyan, Fig. 2b). Arg107 was shown to be important for FMN binding. Mutation of this residue results in a decrease in FMN binding affinity, but it has no effect on C(4a)–peroxyflavin formation and stabilization [52]. Glu175 was found to undergo a conformational change upon binding of FMN and exists in two conformations [12].

Assembly of the enzyme subunits is mediated by interactions among the four-helix bundles, which consist of mostly conserved residues. This subunit tethering involves hydrophobic interactions and the formation of hydrogen bonds along the interface [12, 24]. Tyr151 of the β-subunit was shown to be the key residue required for inter-subunit interaction with the α-subunit. This residue interacts with Phe272 in the loop region of the α-subunit. The mutations βTyr151Lys and βTyr151Trp result in a loss in quantum yield without alteration of subunit dissociation [12].

Bacterial luciferase is protease-labile, possibly due to the presence of a flexible and exposed loop in the enzyme structure. The binding of FMN and phosphate was found to protect the enzyme from proteolytic degradation [34]. Circular dichroism spectroscopic studies indicated a conformational change around the mobile loop portion [35]. The mobile loop connecting the β7 strand and the α7 helix consists of residues between 262 and 291. This segment is highly conserved among α-subunits and is not found in the β-subunit [10]. A loop region located adjacent to the active site cavity was postulated to be important for intermediate stabilization by acting as an active site gate to protect the flavin intermediate from the bulk solvent [10]. However, due to the high flexibility of the loop, the segment between residues 283–290 was not observed in the crystal structure (dashed line in Fig. 2a). The movements of this mobile loop in the presence and absence of $FMNH_2$ were investigated using molecular dynamics simulations. The results indicate the closed conformation of the mobile loop upon $FMNH_2$ binding [11].

3 Mechanism of the Bacterial Luciferase Reaction

Bacterial luciferase is a two-component flavin-dependent monooxygenase that requires a reductase to generate $FMNH^-$ as a substrate for an oxygenase (luciferase) reaction. The flavin is first reduced by NAD(P)H on the flavin reductase, and then transferred to the luciferase. The reduced flavin bound to the luciferase active site reacts with molecular oxygen to form a reactive intermediate, a C4a-oxygenflavin adduct, which is a key intermediate for carrying out monooxygenation [30, 58]. The protonation state of the C4a-oxygenflavin adduct allows different enzymes to perform different reactions. Historically, the adduct obtained from the bacterial luciferase reaction was the first to be investigated by ^{13}C nuclear magnetic resonance (NMR) spectroscopy and the determination that the intermediate is a C4a-oxygen adduct has contributed to the understanding of monooxygenation chemistry by these flavin-dependent enzymes [29]. For enzymes that require flavin to act as a nucleophile, such as bacterial luciferase or Baeyer–Villiger monooxygenases [5], the C4a-oxygen adduct is presumably C4a-peroxyflavin. For electrophilic flavin-dependent monooxygenases such as *p*-hydroxyphenyl acetate hydroxylase [69] or *p*-hydroxybenzoate hydroxylase [17, 22], the intermediate is C4a-hydroperoxyflavin. In both cases, the resulting intermediate after the oxygenation reaction is C4a-hydroxyflavin, which subsequently eliminates water to yield oxidized flavin. In the absence of substrate or under certain conditions, the C4a-oxygen adduct intermediate eliminates hydrogen peroxide directly without performing oxygenation to yield an oxidized flavin, and the stability of the intermediate varies among different enzymes [13, 57, 73].

3.1 Reaction Kinetics of Luciferase With Oxygen in the Presence of Aldehyde

The biochemical reactions and kinetics of bacterial luciferases have been studied extensively [3, 26, 29, 50, 62]. The overall reaction can be summarized by the scheme shown in Fig. 3. $FMNH^-$ is the first substrate that binds to the luciferase to form the enzyme–substrate complex (Lux:$FMNH^-$). Complex formation is followed by a reaction with molecular oxygen to presumably form a C4a-peroxyflavin intermediate (Lux:$FMNHOO^-$). C4a-peroxyflavin reacts with an aldehyde substrate via nucleophilic attack to form a flavin-C4a-peroxyhemiacetal adduct (Lux:FMN-HOOR), which then decays to a carboxylic acid and the excited state enzyme-bound luminescence emitter (C4a-hydroxyflavin, Lux:FMNHOH), which emits blue-green light. The C4a-hydroxyflavin eliminates water and yields oxidized FMN. Alternatively, in the absence of aldehyde, the C(4a)-peroxyflavin is probably protonated and decays to H_2O_2 and FMN without light emission through the dark pathway (Fig. 3). In addition to the reaction in Fig. 3, bacterial luciferases from *P. phosphoreum* and *V. fischeri* can also catalyze a Baeyer–Villiger reaction, as they were reported to incorporate an oxygen atom into 2-tridecanone to yield dodecanoic acid and catalyze the transformation of cyclic ketones [74].

Although the reaction shown in Fig. 3 depicts the luciferase reaction as a sequential order mechanism in which $FMNH^-$ binds to the luciferase first before the reaction with oxygen occurs, followed by the aldehyde reaction. The enzyme can also randomly bind $FMNH^-$ and the aldehyde when both substrates are present [36]. However, the binding of aldehyde prior to $FMNH^-$ can inhibit the luciferase reaction by blocking the subsequent binding of $FMNH^-$ to the enzyme [2, 14, 25]. This inactivation process is reversible, as the inhibitory effect can be removed by diluting an aldehyde-luciferase into an aldehyde-free solution [14, 36]. Incubation of aldehyde with luciferase should be avoided because it results in a lower amount of an active enzyme available for the light-producing reaction (See Sect. 3.4).

The reaction of the luciferase:$FMNH^-$ complex and oxygen is very rapid ($k_2 \sim 2 \times 10^6\ M^{-1}\ s^{-1}$, 4 °C) [3, 62], whereas the binding of $FMNH^-$ to luciferase prior to the oxygen reaction is slower ($k_1 \sim$ 2–4 $\times 10^4\ M^{-1}\ s^{-1}$, 4 °C) with a K_d

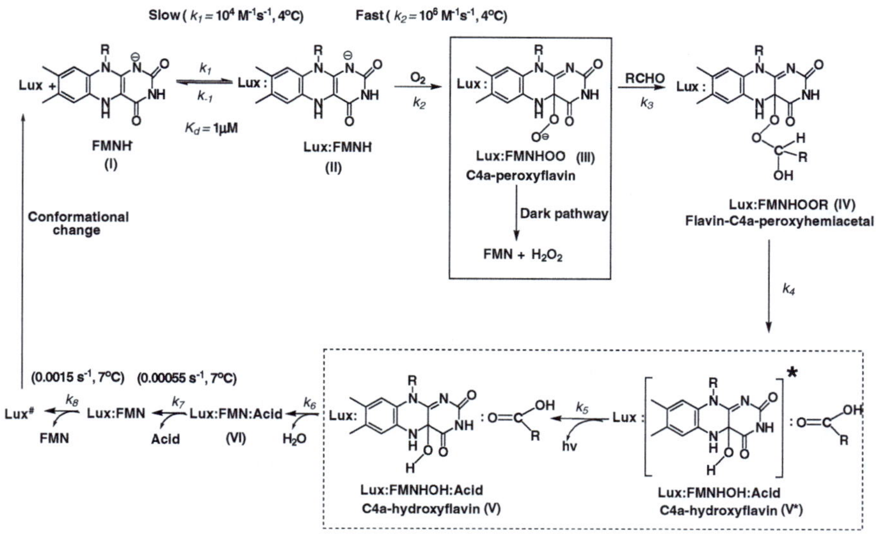

Fig. 3 Overall catalytic mechanism and intermediates of the bacterial luciferase reaction

value in the micromolar range [62, 67]. The reaction of the luciferase:FMNH$^-$ complex and oxygen to form an intermediate C4a-peroxyflavin presumably occurs through a one-electron transfer from flavin to oxygen, resulting in a radical pair that can collapse to form C4a-peroxyflavin that has a maximum absorption peak around 370–380 nm [13, 32, 41]. Under air-saturated conditions, the luciferase:FMNH$^-$ complex reacts with oxygen to form a C4a-peroxyflavin adduct with a rate constant of 350 s^{-1} at 25 °C for the luciferase from *V. harveyi* [3] and 253 s^{-1} at 4 °C for the luciferase from *V. campbellii* [62]. However, the formation of C4a-peroxyflavin is much slower when the luciferase enzyme in an air-saturated solution was mixed with free FMNH$^-$, implying that the binding of free FMNH$^-$ to luciferase is slow and limits the overall reaction [62, 64, 67]. The rate of free FMNH$^-$ binding to luciferase from *V. campbellii* was determined to be around 36.5-fold slower than the oxygen reaction [67].

The C4a-peroxyflavin intermediate in bacterial luciferase is quite stable, allowing it to be isolated and characterized by ^{13}C-NMR spectroscopy. At 2 °C, the half-life of C4a-peroxyflavin is approximately 1 h. The intermediate formed in the luciferase reaction was separated from the other compounds using gel filtration chromatography at −20 °C in the presence of ethylene glycol as an anti-freezing agent [6, 7, 32, 33]. Results of the ^{13}C-NMR spectroscopy indicate that an oxygen atom attaches to the flavin at the C4a position [29]. The luciferase-bound intermediate decays to yield FMN and H_2O_2 in the dark pathway (box in Fig. 3) with an equivalent molar ratio to luciferase-bound C4a-peroxyflavin [33, 70]. If the aldehyde substrate is added into a solution of the purified luciferase-bound C4a-peroxyflavin, the reaction results in light emission and the production of one molecule of FMN per one molecule of the intermediate. The C4a-peroxyflavin has an absorption maximum of approximately

370–380 nm, with a relatively weak fluorescence emission peak of approximately 480–510 nm when the enzyme is excited at 370 nm [6].

An $^{18}O_2$ labeling experiment has shown that one ^{18}O atom from the oxygen molecule is incorporated into the final carboxylic acid product [65]. This result implies that the oxygenation proceeds through the formation of an adduct between the C4a-peroxyflavin and the aldehyde during luciferase catalysis. Although the C4a-peroxyhemiacetal intermediate has not yet been isolated and characterized, spectra of species that may be related to this intermediate were detected by transient kinetics. Macheroux and colleagues [50] have demonstrated that after mixing the isolated C4a-peroxyflavin (III) with slow-reacting aldehydes (dodecanal or octanal), species that may be related to the luciferase bound C4a-peroxyhemiacetal intermediate (IV) were obtained. They observed three sets of isosbestic points and kinetic phases over the course of the change of C4a-peroxyflavin (III) to oxidized FMN (VI). The rates of the second phases are similar to the rate of luminescence decay and showed a 1-[^{2}H] aldehyde isotope effect of approximately 1.5. Therefore, this step was proposed to correspond to the step in which the C(1) aldehyde bond of C4a-peroxyhemiacetal (IV) is broken to liberate an acid and a light-emitting species (k_4). Because the rate of the final phase is slower than the kinetics of light emission decay, this step was assigned as the dehydration step from C4a-hydroxyflavin (V) to form the luciferase bound FMN (VI) (k_6) [41].

The existence of C4a-hydroxyflavin after decomposition of C4a-peroxyhemiacetal was confirmed by the isolation of the intermediate species. Luciferase-bound C4a-hydroxyflavin was isolated by eluting an aerobic buffer through a column containing the anaerobic mixture of luciferase-$FMNH^-$ and decanal [41, 42]. At a higher pH, C4a-hydroxyflavin is more stable than at lower pH because the yield of an intermediate isolated at pH 8.5 was 3.5-fold higher than at pH 7.0. The isolated C4a-hydroxyflavin has a half-life at 2 °C of 33 min with a maximum absorption peak at 360 nm, which is a shorter wavelength than that of C4a-peroxyflavin (370–380 nm). The isolated intermediate has a distinctive fluorescence emission peak at 490 nm [42], which is similar to the spectra of C4a-hydroxyflavin found in other monooxygenases. As the fluorescence emission spectrum of C4a-hydroxyflavin is similar to the luminescence spectrum of the luciferase reaction, it is currently believed that the excited C4a-hydroxyflavin resulting from the C4a-peroxyhemiacetal is the light-emitting species in the luciferase reaction [41, 42, 44].

After the C4a-peroxyhemiacetal (IV) decomposes to yield carboxylic acid and the excited state C4a-hydroxyflavin, light emission is detected, possibly due to the return of the excited C4a-hydroxyflavin (V*) to the ground-state species (V). The mechanism by which the excited C4a-hydroxyflavin is generated is not well understood (dashed box in Fig. 3). However, the formation of C4a-hydroxyflavin (V) was found to occur in parallel with light decay kinetics [41]. After the addition of decanal to isolated C4a-peroxyflavin-bound luciferase, an increase in absorbance at 355 nm, which represents the formation of C4a-hydroxyflavin (V), occurs with a rate constant of 0.01 s^{-1} at 1 °C. The following phase is characterized by an increase in absorption at 440 nm, which represents the dehydration of C4a-hydroxyflavin to FMN (k_6). This process proceeds at a slower rate of 0.0065 s^{-1} at 1 °C [41].

During this step, the carboxylic acid is proposed to still be bound to the luciferase [46]. The dissociation of the ternary complex of luciferase:FMN:acid occurs in an ordered manner, in which release of the carboxylic acid is proposed to occur before dissociation of the flavin. It was found that the FMN dissociation rate was approximately 10–15-fold slower than the light decay rate [46]. The dissociation of FMN from the luciferase:FMN complex as monitored by the change in FMN fluorescence indicates that its kinetics are much slower when the acid is present. The rate constant of the luciferase:FMN:myristic complex dissociation is 0.00055 s^{-1}at 7 °C (k_7), while that of the binary complex of luciferase:FMN is 0.0015 s^{-1} at 7 °C (k_8) [46].

After finishing all oxygenation and flavin oxidation steps, the enzyme releases oxidized FMN very slowly, which is different from the binding kinetics of oxidized FMN to the apoenzyme [1, 62]. Thus, the release of oxidized FMN and the return of the enzyme to an active conformation is thought to involve a change in the conformation of the enzyme. AbouKhair and coworkers [1] investigated the change in luciferase conformation by using chemical modification and probing the proteolytic susceptibility of free luciferase after rapid removal of FMN. The susceptibility of luciferase to modification by an alkylating agent at the reactive thiol of Cys106 in the active site and the protease degradation patterns of luciferase were used in order to monitor the active conformation of the enzyme. The results indicated that after releasing FMN, the half-time for luciferase to return to its active conformation is 25 min at 0 °C (0.00046 s^{-1}). This conformational change that switches the enzyme to the active state is probably one of the rate-limiting steps in the luciferase reaction in addition to the acid release step. Due to the long regeneration time before the luciferase can be active again, most in vitro luciferase assays undergo only a single catalytic turnover because the $FMNH^-$ is usually auto-oxidized before the second cycle of the luciferase reaction takes place [1, 46].

3.2 Reaction Kinetics of Luciferase With Oxygen in the Absence of Aldehyde

The transient kinetics of luciferase-bound $FMNH^-$ and oxygen were studied by stopped-flow spectrophotometry. The kinetics of bacterial luciferases from different sources are varied. For bacterial luciferase from *V. campbellii* under air-saturation at 4 °C, the luciferase-bound $FMNH^-$ was found to react with oxygen rapidly (253 s^{-1}), with a linear dependency on the oxygen concentration. In the absence of aldehyde, the structural rearrangement of the enzyme was evidenced by a change in the C4a-peroxyflavin extinction coefficient (solid box in Fig. 4) before the intermediates eliminate hydrogen peroxide to yield oxidized FMN (Fig. 4) [62]. The decay of C4a-peroxyflavin can be monitored by an increase in absorbance at 450 nm, which reflects the formation of oxidized FMN. The results showed biphasic kinetics, corresponding to the elimination of H_2O_2 at a rate constant of 0.002 s^{-1} and the release of FMN from the luciferase:FMN complex at a rate

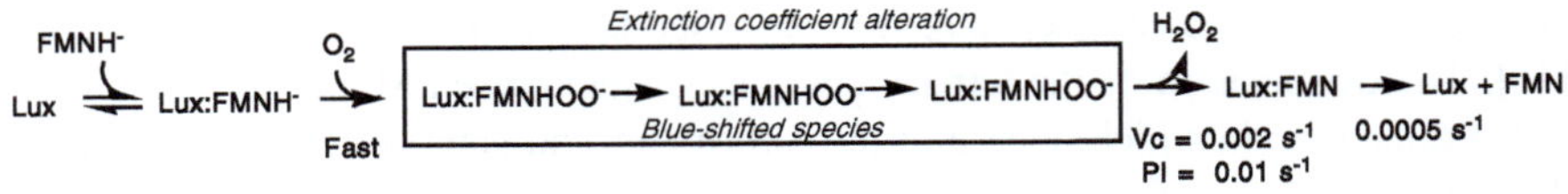

Fig. 4 Kinetic mechanism of the reaction of luciferase-bound $FMNH^-$ with oxygen in the absence of aldehyde; Vc = *V. campbellii* luciferase; Pl = *P. leiognathi* TH1 luciferase [62, 67]

constant of 0.0005 s^{-1} at 4 °C. Therefore, the release of FMN to regenerate the free enzyme is the slowest and rate-limiting step, similar to what was observed in the presence of aldehyde.

The formation of C4a-peroxyflavin in luciferase from *P. leiognathi* TH1, which is categorized as a fast luminescence decay enzyme, is significantly slower than the C4a-peroxyflavin formation in *V. campbellii* luciferase, which is a slow decay type. The reaction of *P. leiognathi* TH1 luciferase with oxygen is also more complicated compared to *V. campbellii* luciferase because there are two phases involved in the formation of C4a-peroxyflavin. Under air-saturation, the faster phase of intermediate formation occurs during the dead time of stopped-flow mixing, while the slow phase occurs with a rate of around 17 s^{-1} at 4 °C [67]. However, the decay of the C4a-peroxyflavin intermediate in *P. leiognathi* TH1 luciferase is much more rapid ($\sim$0.01 s^{-1} at 4 °C) than that of *V. campbellii* luciferase (0.002 s^{-1} at 4 °C) [62, 67, 68]. For *P. leiognathi* TH1 luciferase, the release of FMN from the luciferase:FMN complex is presumably fast because this step could not be detected [67].

3.3 Light Emission Kinetics

The activity of bacterial luciferase is typically assayed by direct injection of $FMNH^-$ into a solution of luciferase or by coupling the reaction with flavin reductase. In the $FMNH^-$ injection assay, FMN is chemically reduced in an anaerobic buffer, and the luciferase reaction is initiated by injection of $FMNH^-$. In the reductase coupling assay, $FMNH^-$ is generated by the flavin reductase. After the reaction is initiated, an initial increase of light emission can be observed, which later decays according to first-order kinetics (Fig. 5). The maximum intensity of light detected at the initial part of the reaction is proportional to the luciferase concentration and directly related to the enzyme activity. The total peak area beneath the light traces represents the total number of emitted photons (quantum yield). The light intensity and decay kinetics vary among luciferases from different origins. These properties are used for enzyme classification into two types—those with fast decay kinetics and those with slow decay kinetics. Luciferases from the genus *Vibrio* and *Photorhabdus* sp. display slow decay kinetics [62, 66, 68], while those from the genus *Photobacteruim* sp. show fast decay [37, 55] (Fig. 5). The luciferase from *P. leiognathi* TH1 also has the distinct characteristic of having higher light intensity than the other luciferases [55].

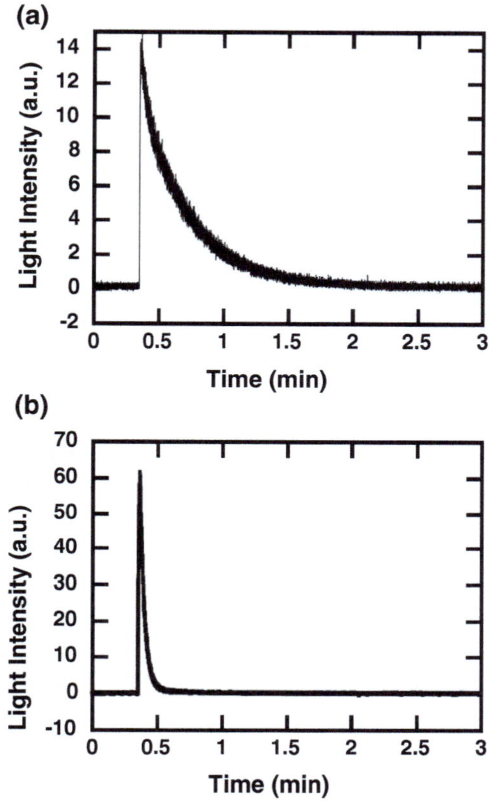

Fig. 5 Light emission kinetics of slow and fast decay luciferases. **a** Slow luminescence decay luciferase from *V. campbellii*. **b** Fast luminescence decay luciferase from *P. leiognathi* TH1. Slow decay luciferases are generally associated with low light intensity, whereas fast decay enzymes are usually associated with high light intensity

The luminescence kinetics also depends on other factors, such as the chain length of the aldehyde (Fig. 6). Using decanal (C-10 aldedyde) as a substrate instead of dodecanal (C-12 aldehyde), *V. campbellii* luciferase changes from having low-intensity and slow-decay luminescence to high-intensity and fast-decay luminescence (unpublished data). Although these two aldehydes give different levels of light intensity and decay kinetics, the total peak area beneath the light traces, which represents the quantum yield, is nearly comparable [25, 43]. The mechanism underlying the influence of aldehyde on the luminescence decay kinetics is not well understood. It was speculated that variation of the aldehyde chain length may affect the steps involved in generation of the light-emitting species, influencing the rates of formation and decay of the C4a-peroxyhemiacetal intermediate (IV). Conversion from slow-decay to fast-decay kinetics in luciferase also occurred when a crucial segment near the luciferase active site was changed (See Sect. 4.3) [37, 48]. The light emission kinetics monitored by stopped-flow spectrophotometry using dodecanal as a substrate at 4 °C indicated that, in some cases, light emission is not homogenous [62]. Two distinct phases of light emission in the *V. campbellii* and *V. harveyi* reactions were detected. The faster light emission phase accounts for 1–2 % of the total light emitted, and it is followed by a major phase of light emission. However,

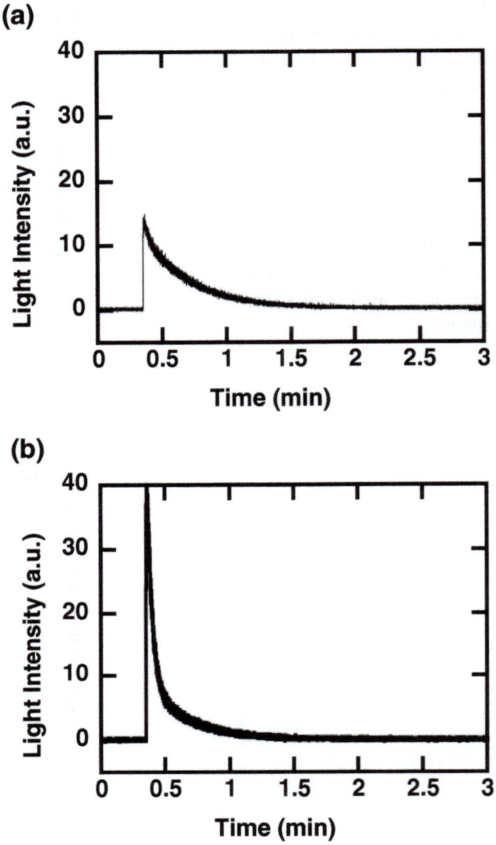

Fig. 6 Effects of aldehyde chain length of **a** dodecanal and **b** decanal on bioluminescence kinetics of *V. campbellii* luciferase reaction

the faster light emission phase disappears at higher temperatures. Therefore, we think that this phenomenon may be caused by a small fraction of enzyme that can catalyze the reaction more rapidly than the majority of the enzyme population, and the ratio of fast- to slow-reacting enzymes in a mixture changes with temperature.

3.4 Inhibition of Luciferase by Aldehyde

Several studies have investigated the inhibition of luciferase by aldehyde and various models of inhibition have been proposed. Some of the studies have proposed that the binding of aldehyde to free luciferase forms a dead-end complex that blocks $FMNH^-$ binding (Fig. 7) [3, 14, 25]. This is based on the results showing that all of the free $FMNH^-$ quickly oxidized back to FMN when $FMNH^-$ was injected into a pre-incubated aerobic solution of luciferase and aldehyde, whereas injection of an aerobic solution of aldehyde into a solution of enzyme and $FMNH^-$ results in the usual formation of the C4a-peroxyflavin intermediate [25]. The inhibition by aldehyde can

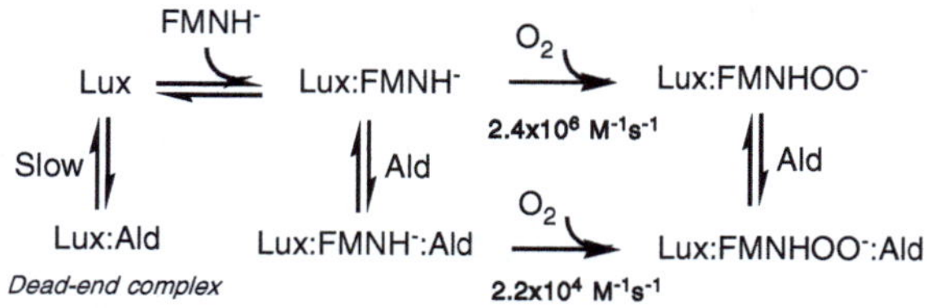

Fig. 7 Aldehyde inhibition caused by the formation of an enzyme-aldehyde dead-end complex and the effect of aldehyde on the oxygen reaction [14, 25]

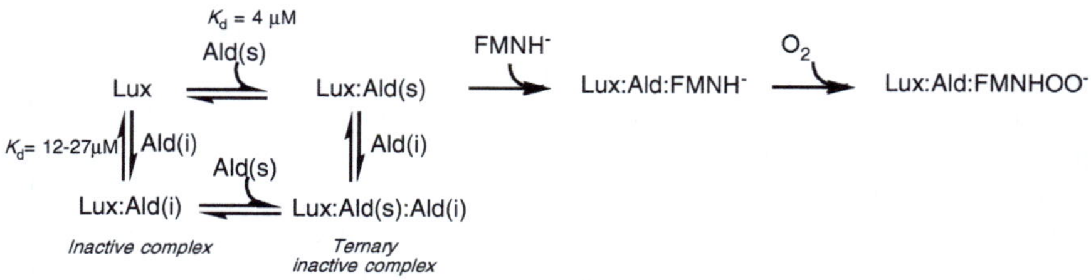

Fig. 8 Inhibition caused by aldehyde binding at two independent sites [43]

be removed if the enzyme solution is diluted, indicating that the inhibition is not caused by covalent modification of the enzyme, or by enzyme denaturation [36]. Although the enzyme:aldehyde complex can dissociate, its decay rate is rather slow [14, 36]. Therefore, $FMNH^-$ reacts with oxygen to yield oxidized FMN before the enzyme:aldehyde complex can dissociate, resulting in aldehyde inhibition. These results suggest that binding of aldehyde may prevent $FMNH^-$ accessibility. The potency of inhibition also increases with aldehyde chain length [25]. Excess aldehyde can also affect the formation of C4a-peroxyflavin and the decay of the enzyme: C4a-hydroxyflavin complex. While the *V. harveyi* luciferase:$FMNH^-$ complex reacts with oxygen to form the C4a-peroxyflavin intermediate at a rate constant of $2.4 \times 10^6\ M^{-1}\ s^{-1}$ [3], the luciferase:$FMNH^-$:decanal complex reacts with oxygen at a rate constant of $2.2 \times 10^4\ M^{-1}\ s^{-1}$ [25].

The model proposing the formation of an aldehyde-enzyme dead-end complex that can prevent $FMNH^-$ binding cannot, however, explain why at low concentrations of aldehyde, no inhibitory effect is observed. This has invoked another model that involves sequential binding of two aldehyde molecules to luciferase [36]. Lei and co-workers [43] have proposed that the aldehyde may bind to two independent binding sites that are different in binding properties and functions (Fig. 8) [43]. One site may have a higher binding affinity and causes no inhibitory effect (substrate site), whereas the other site has lower affinity and the inhibition is more pronounced only at high concentrations of aldehyde (inhibitor site). At low and high (5.3 and 53 μM decanal) concentrations of aldehyde, the aldehyde binding stoichiometry was found to be 0.7 and 1.7 molecules per luciferase [43].

3.5 Mechanism of Light Emitter Generation

Although the light-emitting reaction of luciferase has long been a fascinating area of study, the reaction mechanism that leads to the generation of the excited light-emitting species is still under debate. C4a-hydroxyflavin is believed to be a light-emitting species in the luciferase reaction because its fluorescence characteristics are similar to the emission spectra found in in vivo and in vitro luciferase reactions [41, 42, 44]. Experiments using the C4a-hydroxyflavin analogue, 5-decyl-4a-hydroxy-4a,5-dihydroFMN, to form a complex with *V. harveyi* bacterial luciferase also suggest that the excited state of C4a-hydroxyflavin is a light-emitting species. Although the fluorescence of 5-decyl-4a-hydroxy-4a,5-dihydroFMN is very weak, the fluorescence emission signal at 430 nm of 5-decyl-4a-hydroxy-4a,5-dihydroFMN is increased upon binding to *V. harveyi* luciferase. This result implies that the active site environment of luciferase can enhance light emission of the excited C4a-hydroxyflavin [44].

Several models have been proposed to describe the mechanism of generation of the excited state of C4a-hydroxyflavin, including through a Baeyer–Villiger reaction [19], formation of a dioxirane intermediate [59], and chemically initiated electron exchange luminescence (CIEEL) mechanisms [20]. These models differ mainly with regards to the mechanism of breakdown of the C4a-peroxyhemiacetal, which leads to the generation of the excited light emitter.

The Baeyer–Villiger and dioxirane intermediate mechanisms are similar in the pathway of C4a-peroxyhemiacetal breakdown, but they differ in the identity of the final products proposed. For the Baeyer–Villiger type mechanism, the step involving breakage of the O–O bond in the C4a-peroxyhemiacetal is proposed to be initiated by delocalization of an electron from the oxygen atom of the C1 aldehyde (Fig. 9), resulting in C4a-hydroxyflavin and acid. For the dioxirane mechanism, the rearrangement results in C4a-hydroxyflavin and dioxirane intermediates.

Based on the results from FMN analogue experiments [20, 27] and an observed kinetic isotope effect value of 1.5–1.9 with a 1-[^{2}H] aldehyde substrate [26, 50], the mechanism of formation of the excited state C4a-hydroxyflavin was proposed to occur via the CIEEL mechanism, in which the flavin ring acts as an electron donor to initiate breakage of the O–O bond, leading to the generation of a radical anion-cation pair (Fig. 9). In this mechanism, the excited light emitter is presumably generated from one electron reduction of the C4a-hydroxyflavin radical. Kaaret and Bruice [40] have used the C4a-hydroxyflavin analogue, N(5)-ethyl-4a-hydroxy-4a,5-dihydroluminflavin, as a model to mimic the one electron reduction reaction in the CIEEL mechanism. They found that the chemiluminescence signal could be increased by one-electron electrochemical reduction of the analogue. As substitution of the electron withdrawing group in the FMN molecule results in a decrease in the rate of light decay [20, 27], these researchers proposed that the Baeyer–Villiger type mechanism is not compatible with the results of the experiments with 8-substituted FMNs because the reactions of luciferase with the flavin derivatives

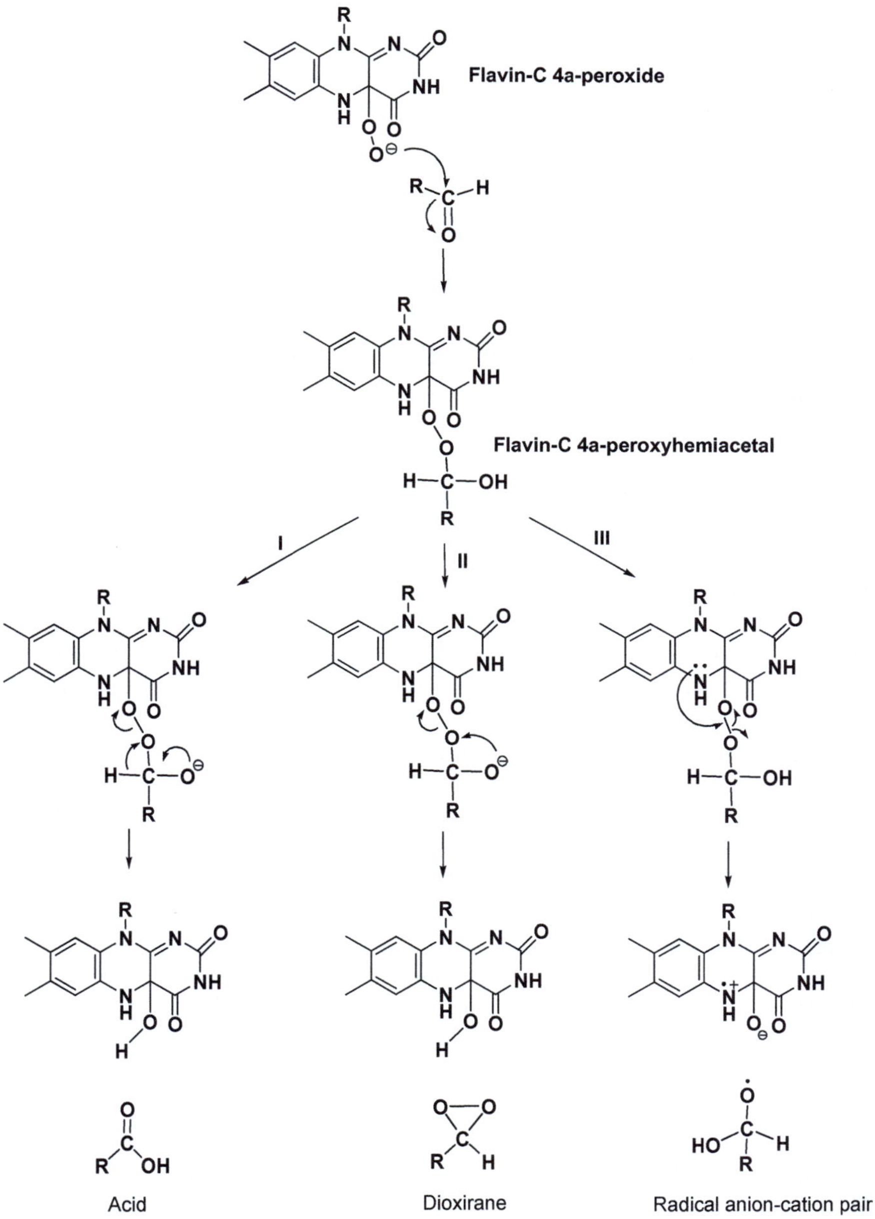

Fig. 9 Three proposed mechanisms related to generation of the excited C4a-hydroxyflavin in the luciferase reaction: **I** Baeyer–Villiger reaction, **II** dioxirane intermediate mechanism, and **III** chemically initiated electron exchange luminescence (CIEEL) or radical anion–cation pair mechanisms

with higher reduction potentials do not display faster decay of bioluminescence as expected, but rather showed slower decay kinetics.

In our opinion, we believe that the effects of the electron withdrawing and donating substituents on the flavin ring are very complicated and can affect the kinetics of many steps, such as generation of the excited light-emitting species and intermediate decay. The existing results do not conclusively rule out the Baeyer–Villiger mechanism, and the mechanism of the light-emitter generation is still unknown and requires further investigation.

3.6 Transfer of Reduced Flavin Between Flavin Reductase and Luciferase

Bacterial luciferase is classified as a two-component flavoprotein monoxygenase [39, 73], in which the reductive and oxidative half-reactions of flavin occur in different protein components. The reduction of FMN is carried out by a reductase and the $FMNH^-$ is subsequently oxidized as a substrate by the oxygenase enzyme. Therefore, in the system, $FMNH^-$ needs to be transferred across the protein components. Because $FMNH^-$ is oxygen-sensitive, during the process of $FMNH^-$ transfer, oxygen exposure must be minimized in order to avoid wasteful $FMNH^-$ oxidation and generation of H_2O_2 as a side-product that is harmful to cells. The transfer of $FMNH^-$ can be conducted either by free diffusion or by a direct transfer mechanism in which protein–protein interactions are required to mediate the transfer.

The reductase of the bacterial luciferase system is presumably encoded by the *luxG* gene, and the luciferase enzyme is encoded by the *luxAB* genes in the *lux* operon (Fig. 1). The reduction of FMN to generate $FMNH^-$ occurs on the LuxG oxidoreductase to supply $FMNH^-$ for the light-producing reaction on the luciferase. Complex formation between LuxG and luciferase could not be detected by gel filtration chromatography [55]. Our group has investigated the mechanism of $FMNH^-$ transfer using transient kinetics because this method can clearly measure the individual rate constants of enzyme reactions [63, 64]. More details about methodology and interpretation have been reviewed elsewhere [64].

The rate constants of the critical steps involved, including $FMNH^-$ transfer, dissociation of $FMNH^-$ from LuxG, and binding of $FMNH^-$ to luciferase, were investigated and compared when only one or both proteins were present [54, 64, 67]. The results showed that the presence of LuxG or luciferase has no influence on the rate constants of the other protein component. The data imply that these two proteins do not require complex formation to transfer $FMNH^-$ [67] and strongly support the free diffusion model, in which $FMNH^-$ dissociates from LuxG and freely diffuses to bind to the luciferase (Fig. 10).

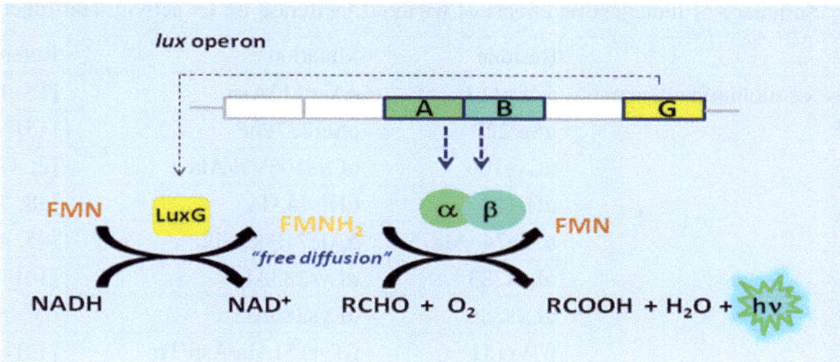

Fig. 10 The transfer of $FMNH^-$ between LuxG oxidoreductase and luciferase occurs via a free diffusion mechanism

4 Mutagenesis Studies of Bacterial Luciferase

Bacterial luciferase variants were first generated in luminous bacteria (*V. harveyi*) through chemical-induced mutagenesis. Various phenotypes including dim, red-shifted, and temperature-sensitive mutants were obtained [15]. Later on, when the protein sequence and structural information became available, the roles of particular residues were investigated by performing a series of chemical modifications and site-directed mutagenesis. A number of residues were found to be important for the structural and catalytic functions of bacterial luciferase. Most of these residues reside in the α-subunit, while a few residues in the β-subunit also contribute indirectly to the structural integrity of the enzyme. A summary of adverse effects resulting from the change of particular amino acids in bacterial luciferase is listed in Table 1. The positions of these residues in the luciferase structure are also shown in Fig. 11.

4.1 Variants with Decreased Bioluminescence Activity

Dark or dim variants are the enzymes that are impaired in light production. Dark variants of *V. harveyi* luciferase obtained in early studies were generated by chemical mutagenesis. All of the mutations were located in the α-subunit [15]. The sequence of some of the well-characterized mutants, such as AK-6 and AK-20, were later analyzed and the mutations were identified as αAsp113Asn and αSer227Phe, respectively [16]. The AK-6 mutant is impaired in flavin binding affinity, while the aldehyde binding of the AK-20 mutant is defective. Based on the crystal structure, αAsp113 is located close to the pyrimidine portion on the *si*-face of the flavin ring. This residue forms part of the cavity wall close to the β-subunit interface. Oδ2 of αAsp113 is 6 Å away from the N3 of the isoalloxazine ring (Fig. 12a) and can potentially form electrostatic interactions with neighboring

Table 1 Summary of mutagenesis effects of bacterial luciferase on its activity and function

Effects	Residue	Mutation	References
Decrease of bioluminescence activity	αAsp113	αAsp113Asn	[15, 16]
	αSer227	αSer227Phe	[15]
	αCys106	αCys106Val/Ala	[2, 47, 76]
	αHis44	αHis44Ala	[38, 77]
	αAla74–Ala75	αAla74Phe/Gly	[45, 47]
	αLys283	αLys283Ala	[10]
	αLys286	αLys286Ala	
	βTyr151	βTyr151Ala/Asp/Trp	[12]
Shift of light emission color	αVal173	αVal173Ala/Cys/Thr/Ser	[48]
	αCys106	αCys106Val	[47]
	αAla75	αAla75Gly	
	αAsp113	αAsp113Asn	[16]
Increase of bioluminescence decay rate	αGlu175	αGlu175Gly	[37]
Decrease of C4a-peroxyflavin intermediate stability	αGlu175	αGlu175Gly	[37]
	αCys106	αCys106Val/Ala	[3, 47, 76]
	αLys283	αLys283Ala	[10]
Ability to express in eukaryotic cells	–	Fusion of α- and β-sub units with linker peptide	[8, 23, 56, 68, 75]

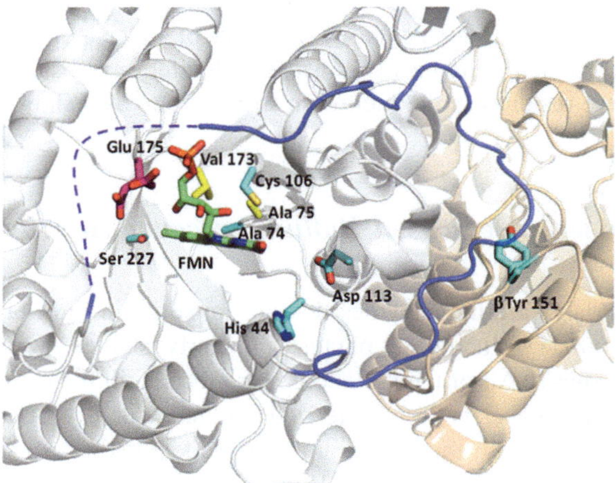

Fig. 11 The structure of bacterial luciferase from *V. harveyi* (pdb: 3FGC) with the side chains shown for residues investigated by mutagenesis. Key residues that are crucial for bioluminescence activity are colored *cyan*. *Yellow* and *magenta* represent the residues important for controlling light emission color and light decay kinetics, respectively. The αLys283 and αLys286 that are located in the unstructured loop region (*dashed line*) cannot be seen in this structure

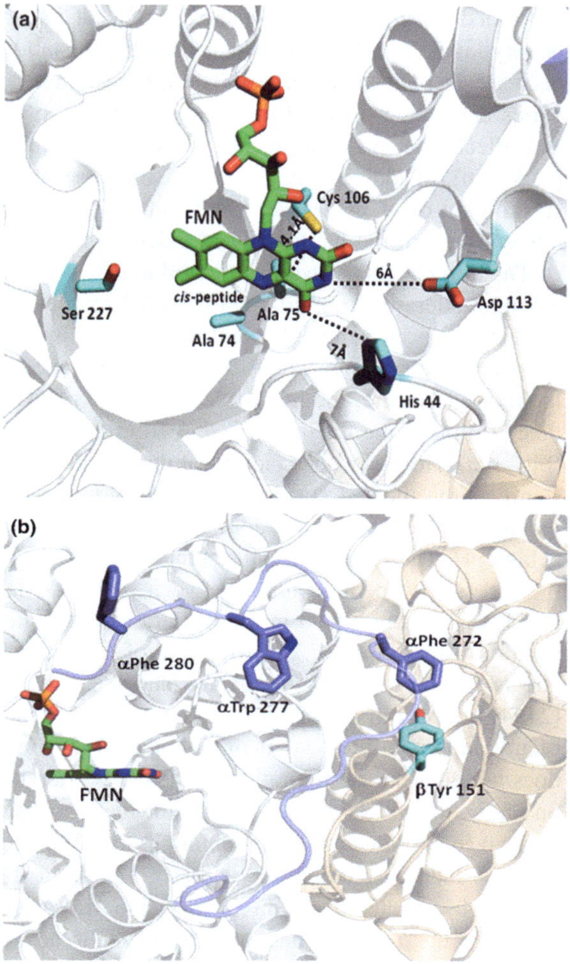

Fig. 12 **a** The active site of bacterial luciferase with FMN bound. The αCys106 and the *cis*-peptide bond of Ala74-Ala75 are located on the *re*-face side of isoalloxazine ring. αAsp113 and αHis 44 are located close to the pyrimidine portion of the flavin ring on the *si*-face side. αSer227, which is important for aldehyde binding, is located near the benzenoid part. **b** Interactions around the subunit interface between Tyr151 of the β-subunit and the hydrophobic residues in the mobile loop portion of the α-subunit

residues αHis44, αLys112, and αHis45, which may be important for catalytic function [12]. Therefore, mutation of αAsp113 may perturb the electrostatic properties around this reaction site. The residue αSer227 is located near the benzenoid moiety of the isoalloxazine ring, which is in a highly hydrophobic part of the active site cavity postulated to be the aldehyde binding site.

Using chemical modification techniques, the results revealed a role of the reactive thiol group of the αCys106 side chain in the luciferase reaction [53]. Alkylation of the thiol group of αCys106 had previously been shown to inactivate the enzymatic reaction and block flavin and aldehyde binding [28]. αCys106 was also found to be important for the stability of C4a-peroxyflavin. According to the bacterial luciferase structure, αCys106 is located near the *re*-face of the flavin ring and is 4.1 Å away from the C4a position of the isoalloxazine ring (Fig. 12a) [12]. The αCys106 reactive thiol points toward the C4a position where a peroxide adduct is formed during luciferase catalysis. The αCys106 variants can still form C4a-peroxyflavin, but the mutation decreases its stability and ultimately results in a decrease in light yield [3, 76]. The luminescence output of the αCys106 mutants varies depending on what the residue is changed to. Substitution of the Cys with Val affects the bioluminescence activity more than substitution with Ala or Ser [4, 47, 76]. In addition, αCys106 is also important for the binding of aldehyde. The substitution of αCys106 with Ser resulted in a 10-fold increase in the dissociation rate constant of the luciferase and aldehyde complex [3, 43].

The dim phenotype was also found in αHis44 variants. This residue is conserved among luciferases from different bacterial sources. It is crucial for bioluminescence production because the light yield was reduced by six orders of magnitude relative to the wild-type enzyme when αHis44 was mutated to Ala [77]. According to the luciferase structure, the imidazole side chain of αHis44 is located near the pyrimidine portion of the flavin and is 7 Å from the C4-O atom of the isoalloxazine ring (Fig. 12a). However, the stability of the C4a-peroxyflavin in αHis44 variants is not significantly affected compared to the wild-type enzyme. The activity of αHis44 variants can be rescued by adding free imidazole and the luminescence level is high at high pH. These findings suggest that αHis44 may function as a catalytic base [38].

The rare *cis*-peptide bond between αAla74–Ala75 located close to the flavin isoalloxazine ring bound in the luciferase active site is important for the luciferase reaction (Fig. 12a) [12, 24]. Mutation of each αAla74 or αAla75 yielded different effects. While mutation of αAla75 showed a slight decrease in enzyme activity [24, 47], substitution of αAla74 with Gly or Phe resulted in a decrease of light yield of around two to three orders of magnitude [45]. It was speculated that the mutation does not alter the configuration of the *cis*-peptide bond but rather causes the strain on the bond, which may be crucial for luciferase activity.

Two Lys residues in the α-subunit loop region were reported to be important for bacterial luciferase quantum yield [10, 12]. A large decrease in light intensity was found when either αLys283 or αLys286 was mutated to Ala. The mobile loop portion of the α-subunit has been speculated to act as a gate to protect intermediates in the luciferase reaction [49]. Deletion of the whole loop caused a big loss in the reaction quantum yield without significant alteration in carboxylic acid production (oxygenation) efficiency [61]. The substantial decrease in light production of the Lys283Ala mutant was found to be due to the defect in C4a-peroxyflavin intermediate stabilization [10].

Some of the residues located in the β-subunit are also important for controlling bioluminescence activity originating from the α-subunit. βTyr151 of the β-subunit is part of the inter-subunit interactions that are crucial for enzyme activity. Although it has long been known that the β-subunit helps in maintaining the structural integrity of the α-subunit, the specific roles of these β-subunit amino acids are not well understood [78]. βTyr151 was identified as one of the residues in the β-subunit critical for linking the functional roles of the two subunits. When βTyr151 was mutated to Trp or Asp, the light quantum yield was largely reduced to be less than 1 % compared to the wild-type enzyme [12]. This mutation also resulted in a decrease in flavin binding affinity and the integrity of enzyme structure because the βTyr151 mutants are more prone to heat inactivation. βTyr151 is positioned at the α/β-subunit interface and may have hydrophobic interactions with the residues αPhe272, αTrp277, αPhe280 located in the loop region of the α-subunit (Fig. 12b). These interactions may help in maintaining the active conformation of the α-subunit. Up to now, mutation of βTyr151 is the only mutation in β-subunit shown to significantly affect the function of the α-subunit.

4.2 Variants With Color Shift

One of the most active areas in gene reporter research is in engineering reporter systems to display various colors. Mutation of firefly and beetle luciferases yielded color-shifted variants in which the emission energy is altered by the change in the chromophore binding environment [9]. Several attempts have also been made to construct color-shifted variants in the bacterial luciferase system. However, most of the efforts resulted in a marked decrease in bioluminescence yield with a small shift of λ_{max}. The light emission shift was speculated to be the consequence of altering the dielectric field in the chromophore binding site, which also perturbs the luminescence quantum yield [47].

All of the key residues that control the light color of bacterial luciferases are located around the isoalloxazine ring binding site (Fig. 13) that presumably would interact with the luminescent C4a-hydroxyflavin intermediate. The first red-shifted variant obtained was the AK-6 dark mutant (αAsp113Asn) that displays a 12-nm red-shifted phenotype [15, 16]. For the αCys106Val variant, the shift in the bioluminescence spectrum of approximately 8–10 nm is also accompanied by a large decrease in light yield [47]. Mutation of αVal173, in which its isopropyl side chain is located near the dimethyl benzenoid ring, also perturbs the emission spectrum. Replacement of αVal173 with hydrophobic residues, such as Leu or Ile, do not change the emission color of the enzyme, while mutations to residues having greater polarity resulted in a shift to emissions at longer wavelengths [48].

Making additional amino acid substitutions to the variants that were reported to cause color shifts resulted in variants with longer wavelength shifts in the emission spectrum and improvement of luminescence intensity. The addition of an αAla75Gly mutation on the αCys106Val variant resulted in a double variant that has a smaller emission spectrum shift but a luminescence level greater than the single variant of αCys106Val [47]. The restoration of the luminescence intensity of this variant may be due to the alleviation of the steric effects generated by the αCys106Val mutation when αAla75 is changed to Gly [47]. Further mutation at αVal173 in addition to the αAla75Gly/Cys106Val mutations resulted in a triple mutant enzyme, which displayed a ~ 15-nm shift in its emission spectrum, with a bioluminescent activity close to that of the wild-type enzyme [47, 48].

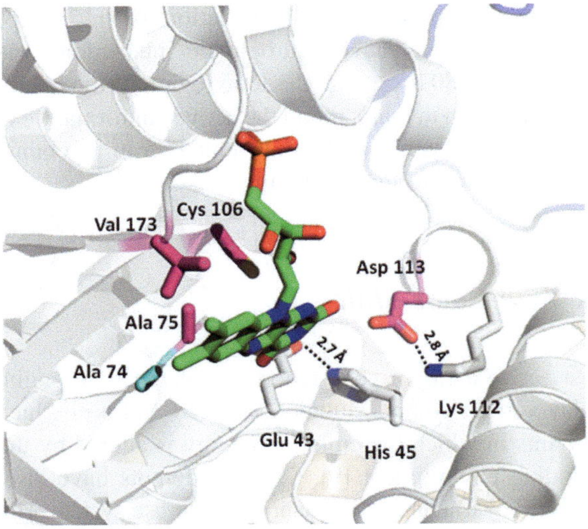

Fig. 13 Residues controlling the light color of bacterial luciferase are located around the isoalloxazine ring binding site. αCys106, αVal173, and αAla75 are close to each other and located along the *re*-face of the isoalloxazine plane. αAsp113 plays a role in controlling the electrostatic properties near the pyrimidine side of the flavin by forming hydrogen bond networks with neighboring residues

4.3 Variants With Alterations in Light Emission Kinetics

As mentioned previously, bacterial luciferases can be classified based on their light decay kinetics into two types: those with fast light decay and those with slow light decay. Although the key factors that control this process are not yet known, results from several experiments have identified some residues that may be responsible for controlling the light kinetics.

Construction of chimeric enzymes by combining fast (*P. phosphoreum*) and slow decay (*P. luminescens*) luciferases has identified the central part of the α-subunit as being an important part in harnessing light. Replacement of a 67 amino acid-length piece of the central region of the α-subunit from *P. phosphoreum* luciferase to the corresponding region of *P. luminescens* luciferase converted the enzyme from the slow to fast decay type [72]. Subsequent studies using random mutagenesis in this region showed that αGlu175 is the critical residue in dictating the light decay kinetics [37]. Based on the luciferase structure, the side chain of αGlu175 is one of several residues that form the FMN 5′-phosphate binding site (Fig. 2b). The residue can also undergo a conformational change upon the binding of FMN [12]. An increase in the light decay rate was found when αGlu175 was mutated to Gly. The stability of the C4a-peroxyflavin of this variant was markedly decreased [37].

We note that the stability of the C4a-peroxyflavin is often, but not always, associated with the light decay kinetics. The decay rates of C4a-peroxyflavin in luciferases from *V. campbellii* and *V. harveyi,* which are slow decay enzymes, were markedly slower than the decay rate found in the fast decay luciferase from *P. leiognathi* TH1 [3, 62, 67, 68]. However, mutations that decrease C4a-peroxyflavin stability, such as αCys106Val, do not result in fast decay kinetics [47]. Thus, it is clear that αCys106 is not involved in the control of the kinetics of light decay, and more residues in addition to αGlu175 that are important for this process need to be identified.

4.4 Fusion Luciferase for Bioreporter Applications

The bioluminescence activity of bacterial luciferase has allowed it to be applied as a reporter gene in prokaryotic and eukaryotic systems. To enable bioreporter applications of bacterial luciferase in eukaryotic cells in which gene expression is required to be monocistronic, construction of fusion *luxA* and *luxB* genes was carried out such that the resulting fusion gene is controlled by a single promoter upstream of the gene [8, 23, 56, 68, 75]. Various peptide linkers from 1 to 22 amino acids were used to link the α- and β-subunits of bacterial luciferases [51].

The bioluminescence activity of the fusion luciferase is strongly dependent on the length and sequence of a linker peptide. The linkers with one or two amino acids showed very low activities, while those with around ten amino acids resulted in much more prominent activities [8, 23, 51]. With a decapeptide linker, purified fusion luciferase from *V. harveyi* exhibited a specific activity of approximately 50–80 % compared to the wild-type enzyme [51]. We have obtained a fusion luciferase with activity almost as high as the wild-type enzyme [68].

Even though most purified fusion enzymes give reasonable activity compared to the wild-type enzyme, very low activity in vivo was observed when the enzyme was expressed around 30 °C and above. This may be due to the higher tendency of the protein-folding problem of fusion enzymes to occur at higher temperatures. The fusion luciferase from *V. harveyi* with a decapeptide linker gave only 0.02 % of the in vivo bioluminescence level of the native luciferase when expressed at 37 °C in *E. coli* [23]. Later on, after a more thermostable luciferase from terrestrial bacteria, *P. luminescens,* was cloned and characterized, the fusion luciferase with a 12-amino acid linker was constructed; this enzyme showed superior characteristics to the corresponding *V. harveyi* enzyme [75]. The fusion enzyme retained more than 20 % of the in vivo activity of the native heterodimeric enzyme when expressed at 37 °C.

Our group constructed a fusion luciferase from *V. campbellii* in which the enzyme is much superior in flavin binding and thermostability as compared to the *V. harveyi* enzyme [62, 68]. The fusion enzyme with a decapeptide linker retained activity as high as 35 % relative to the wild-type enzyme when expressed at 37 °C in *E. coli* [68]. In our opinion, among the bacterial enzymes reported to date, the fusion *V. campbellii* luciferase is the most suitable enzyme for eukaryotic bioreporter applications.

5 Concluding Remarks

Bacterial luciferase is a distinctive two-component flavin-dependent monooxygenase in that the monooxygenation reaction generates concomitant light emission. The incorporation of one oxygen atom from molecular oxygen into an aldehyde is mediated through a reactive flavin C4a-oxygen-adduct, presumably C4a-peroxyflavin. Several lines of evidence suggest that the following intermediate, C4a-hydroxyflavin, is the species responsible for this light emission. Although the overall reaction of bacterial luciferase is understood, many unknown questions remain, including how the enzyme controls the active site environment to generate C4a-peroxyflavin rather than C4a-hydroperoxyflavin to attain maximum reaction efficiency with aldehyde, as well as what chemical mechanism underlies the generating of light emitters.

Bacterial luciferase has promising potential in bioreporter applications. Its current usage is under-exploited, probably due to the requirement for flavin reductase in the reaction and problems in mammalian cell expression. With the current understanding of how luciferase and reductase coordinate their overall reaction and the ability to express the enzyme in mammalian cells, these obstacles should be overcome and the luciferase system will provide both viable and economical gene reporters. However, enhancement or fine-tuning of the enzyme by mutagenesis to increase the quantum yield, alter the color, slow down the light decay kinetics, and improve the protein expression in mammalian cells is needed for further development towards robust gene reporters.

Acknowledgments The authors acknowledge research support from The Thailand Research Fund RTA5680001 (to P.C.), Mahidol University (to R.T. and P.C.) and Mahidol University (Talent Management Project to R.T.). We are grateful to Danaya Pakotiprapha for assistance with PyMOL figure creation.

References

1. AbouKhair NK, Ziegler MM, Baldwin TO (1985) Bacterial luciferase: demonstration of a catalytically competent altered conformational state following a single turnover. Biochemistry 24:3942–3947
2. Abu-Soud HM, Clark AC, Francisco WA, Baldwin TO, Raushel FM (1993) Kinetic destabilization of the hydroperoxy flavin intermediate by site-directed modification of the reactive thiol in bacterial luciferase. J Biol Chem 268:7699–7706
3. Abu-Soud HM, Mullins LS, Baldwin TO, Raushel FM (1992) Stopped-flow kinetic analysis of the bacterial luciferase reaction. Biochemistry 31:3807–3813
4. Baldwin TO, Chen LH, Chlumsky LJ, Devine JH, Ziegler MM (1989) Site-directed mutagenesis of bacterial luciferase: analysis of the essential thiol. J Biolumin Chemilumin 4:40–48
5. Balke K, Kadow M, Mallin H, Saß S, Bornscheuer UT (2012) Discovery, application and protein engineering of Baeyer–Villiger monooxygenases for organic synthesis. Org Biomol Chem 10:6249–6265

6. Balny C, Hastings JW (1975) Fluorescence and bioluminescence of bacterial luciferase intermediates. Biochemistry 14:4719–4723
7. Becvar JE, Tu S-C, Hastings JW (1978) Activity and stability of the luciferase-flavin intermediate. Biochemistry 17:1807–1812
8. Boylan M, Pelletier J, Meighen EA (1989) Fused bacterial luciferase subunits catalyze light emission in eukaryotes and prokaryotes. J Biol Chem 264:1915–1918
9. Branchini BR, Southworth TL, Khattak NF, Michelini E, Roda A (2005) Red- and green-emitting firefly luciferase mutants for bioluminescent reporter applications. Anal Biochem 345:140–148
10. Campbell ZT, Baldwin TO (2009) Two lysine residues in the bacterial luciferase mobile loop stabilize reaction intermediates. J Biol Chem 284:32827–32834
11. Campbell ZT, Baldwin TO, Miyashita O (2010) Analysis of the bacterial luciferase mobile loop by replica-exchange molecular dynamics. Biophys J 99:4012–4019
12. Campbell ZT, Weichsel A, Montfort WR, Baldwin TO (2009) Crystal structure of the bacterial luciferase/flavin complex provides insight into the function of the β subunit. Biochemistry 48:6085–6094
13. Chaiyen P, Fraaije MW, Mattevi A (2012) The enigmatic reaction of flavins with oxygen. Trends Biochem Sci 37:373–380
14. Choi H, Tang C-K, Tu S-C (1995) Catalytically active forms of the individual subunits of *Vibrio harveyi* luciferase and their kinetic and binding properties. J Biol Chem 270:16813–16819
15. Cline TW, Hastings JW (1972) Mutationally altered bacterial luciferase. Implications for subunit functions. Biochemistry 11:3359–3370
16. Cline TW, Hastings JW (1974) Mutated luciferases with altered bioluminescence emission spectra. J Biol Chem 249:4668–4669
17. Cole LJ, Entsch B, Ortiz-Maldonado M, Ballou DP (2005) Properties of *p*-hydroxybenzoate hydroxylase when stabilized in its open conformation. Biochemistry 44:14807–14817
18. Delong EF, Steinhauer D, Israel A, Nealson KH (1987) Isolation of the *lux* gene from *Photobacterium leiognathi* and expression in *Escherichia coli*. Gene 54:203–210
19. Eberhard A, Hastings JW (1972) A postulated mechanism for the bioluminescent oxidation of reduced flavin mononucleotide. Biochem Biophys Res Commun 47:348–353
20. Eckstein JW, Hastings JW, Ghisla S (1993) Mechanism of bacterial bioluminescence: 4a,5-dihydroflavin analogs as models for luciferase hydroperoxide intermediates and the effect of substituents at the 8-position of flavin on luciferase kinetics. Biochemistry 32:404–411
21. Engebrecht J, Simon M, Silverman M (1985) Measuring gene expression with light. Science 227:1345–1347
22. Entsch B, Cole LJ, Ballou DP (2005) Protein dynamics and electrostatics in the function of *p*-hydroxybenzoate hydroxylase. Arch Biochem Biophys 433:297–311
23. Escher A, O'Kane DJ, Lee J, Szalay AA (1989) Bacterial luciferase αβ fusion protein is fully active as a monomer and highly sensitive in vivo to elevated temperature. Proc Nati Acad Sci USA 86:6528–6532
24. Fisher AJ, Thompson TB, Thoden JB, Baldwin TO (1996) The 1.5 Å resolution crystal structure of bacterial luciferase low salt conditions. J Biol Chem 271:21956–21968
25. Francisco WA, Abu-Soud HM, Baldwin TO, Raushel FM (1993) Interaction of bacterial luciferase with aldehyde substrates and inhibitors. J Biol Chem 268:24734–24741
26. Francisco WA, Abu-Soud HM, DelMonte AJ, Singleton DA, Baldwin TO, Raushel FM (1998) Deuterium kinetic isotope effects and the mechanism of the bacterial luciferase reaction. Biochemistry 37:2596–2606
27. Francisco WA, Abu-Soud HM, Topgi R, Baldwin TO, Raushel FM (1996) Interaction of bacterial luciferase with 8-substituted flavin mononucleotide derivatives. J Biol Chem 27:104–110
28. Fried A, Tu S-C (1984) Affinity labeling of the aldehyde site of bacterial luciferase. J Biol Chem 259:10754–10759

29. Ghisla S, Hastings JW, Favaudon V, Lhoste JM (1978) Structure of the oxygen adduct intermediate in the bacterial luciferase reaction: ^{13}C nuclear magnetic resonance determination. Proc Natl Acad Sci USA 75:5860–5863
30. Ghisla S, Massey V (1989) Mechanisms of flavoprotein-catalyzed reactions. Eur J Biochem 181:1–17
31. Gunsalus-Miguel A, Meighen EA, Nicoli MZ, Nealson KH, Hastings JW (1972) Purification and properties of bacterial luciferases. J Biol Chem 247:398–404
32. Hastings JW, Balny C, Peuch CL, Douzou P (1973) Spectral properties of an oxygenated luciferase-flavin intermediate isolated by low-temperature chromatography. Proc Natl Acad Sci USA 70:3468–3472
33. Hastings JW, Balny C (1975) The oxygenated bacterial luciferase-flavin intermediate. Reaction products via the light and dark pathways. J Biol Chem 250:7288–7293
34. Holzman TF, Baldwin TO (1980) Proteolytic inactivation of luciferases from three species of luminous marine bacteria, *Beneckea harveyi*, *Photobacterium fischeri*, and *Photobacterium phosphoreum*: evidence of a conserved structural feature. Proc Natl Acad Sci USA 77:6363–6367
35. Holzman TF, Baldwin TO (1980) The effects of phosphate on the structure and stability of the luciferases from *Beneckea harveyi*, *Photobacterium fischeri*, and *Photobacterium phosphoreum*. Biochem Biophys Res Commun 94:1199–1206
36. Holzman TF, Baldwin TO (1983) Reversible inhibition of the bacterial luciferase catalyzed bioluminescence reaction by aldehyde substrate: kinetic mechanism and ligand effects. Biochemistry 22:2838–2846
37. Hosseinkhani S, Szittner R, Meighen EA (2005) Random mutagenesis of bacterial luciferase: critical role of Glu^{175} in the control of luminescence decay. Biochem J 385:575–580
38. Huang S, Tu S-C (1997) Identification and characterization of a catalytic base in bacterial luciferase by chemical rescue of a dark mutant. Biochemistry 36:14609–14615
39. Huijbers MM, Montersino S, Westphal AH, Tischler D, van Berkel WJ (2014) Flavin dependent monooxygenases. Arch Biochem Biophys 544:2–17
40. Kaaret TW, Bruice TC (1990) Electrochemical luminescence with N(5)-ethyl-4a-hydroxy-3-methyl-4a,5-dihydrolumiflavin. The mechanism of bacterial luciferase. Photochem Photobiol 51:629–633
41. KÜrfurst M, Ghisla S, Hastings JW (1984) Characterization and postulated structure of the primary emitter in the bacterial luciferase reaction. Proc Natl Acad Sci USA 81:2990–2994
42. Kurfuerst M, Macheroux P, Ghisla S, Hastins JW (1987) Isolation and characterization of the transient, luciferase-bound flavin-4a-hydroxide in the bacterial luciferase reaction. Biochim Biophys Acta 924:104–110
43. Lei B, Cho KW, Tu S-C (1994) Mechanism of aldehyde inhibition of *Vibrio harveyi* luciferase. Identification of two aldehyde sites and relationship between aldehyde and flavin binding. J Biol Chem 269:5612–5618
44. Lei B, Ding Q, Tu S-C (2004) Identity of the emitter in the bacterial luciferase luminescence reaction: binding and fluorescence quantum yield studies of 5-decyl-4a-hydroxy-4a,5-dihydroriboflavin-5'-phosphate as a model. Biochemistry 43:15975–15982
45. Li C-H, Tu S-C (2005) Active site hydrophobicity is critical to the bioluminescence activity of *Vibrio harveyi* luciferase. Biochemistry 44:12970–12977
46. Li Z, Meighen EA (1994) The turnover of bacterial luciferase is limited by a slow decomposition of the ternary enzyme-product complex of luciferase, FMN, and fatty acid. J Biol Chem 269:6640–6644
47. Lin LY-C, Sulea T, Szittner R, Kor C, Purisima EO, Meighen EA (2002) Implications of the reactive thiol and the proximal non-proline *cis*-peptide bond in the structure and function of *Vibrio harveyi* luciferase. Biochemistry 41:9938–9945
48. Lin LY-C, Szittner R, Friedman R, Meighen EA (2004) Changes in the kinetics and emission spectrum on mutation of the chromophore-binding platform in *Vibrio harveyi* luciferase. Biochemistry 43:3183–3194

49. Low JC, Tu S-C (2002) Functional roles of conserved residues in the unstructured loop of *Vibrio harveyi* bacterial luciferase. Biochemistry 41:1724–1731
50. Macheroux P, Ghisla S, Hastings JW (1993) Spectral detection of an intermediate preceding the excited state in the bacterial luciferase reaction. Biochemistry 32:14183–14186
51. Meighen EA (1991) Molecular biology of bacterial bioluminescence. Microbiol Mol Biol Rev 55:123–142
52. Moore C, Lei B, Tu S-C (1999) Relationship between the conserved α subunit arginine 107 and effects of phosphate on activity and stability of *Vibrio harveyi* luciferase. Arch Biochem Biophys 370:45–50
53. Nicoli MZ, Meighen EA, Hastings JW (1974) Bacterial luciferase. Chemistry of the reactive sulfhydryl. J Biol Chem 249:2385–2392
54. Nijvipakul S, Ballou DP, Chaiyen P (2010) Reduction kinetics of a flavin oxidoreductase LuxG from *Photobacterium leiognathi* (TH1): Half-sites reactivity. Biochemistry 49:9241–9248
55. Nijvipakul S, Wongratana J, Suadee C, Entsch B, Ballou DP, Chaiyen P (2008) LuxG is a functioning flavin reductase for bacterial luminescence. J Bacteriol 190:1531–1538
56. Olsson O, Escher A, Sandberg G, Schell J, Koncz C, Szalay AA (1989) Engineering of monomeric bacterial luciferases by fusion of *luxA* and *luxB* genes in *Vibrio harveyi*. Genes 81:335–347
57. Palfey BA, Massey V (1998) Flavin-dependent enzymes. In: Sinnott M (ed) Comprehensive biological catalysis. Academic Press, San Deigo, A mechanistic reference, pp 1–27
58. Palfey BA, McDonald CA (2010) Control of catalysis in flavin-dependent monooxygenases. Arch Biochem Biophys 493:26–36
59. Raushel FM, Baldwin TO (1989) Proposed mechanism for the bacterial bioluminescence reaction involving a dioxirane intermediate. Biochem Biophys Res Commun 164:1137–1142
60. Shimomura O (2006) Bioluminescence: chemical principles and methods, 1st edn. World Scientific Publishing Co. Pte. Ltd., Singapore, pp 30–46
61. Sparks JM, Baldwin TO (2001) Functional implications of the unstructured loop in the (β/α)8 barrel structure of the bacterial luciferase α subunit. Biochemistry 40:15436–15443
62. Suadee C, Nijvipakul S, Svasti J, Entsch B, Ballou DP, Chaiyen P (2007) Luciferase from *Vibrio campbellii* is more thermostable and binds reduced FMN better than its homologues. J Biochem 142:539–552
63. Sucharitakul J, Phongsak T, Entsch B, Svasti J, Chaiyen P, Ballou DP (2007) Kinetics of a two-component *p*-hydroxyphenylacetate hydroxylase explain how reduced flavin is transferred from the reductase to the oxygenase. Biochemistry 46(783):8611–8623
64. Sucharitakul J, Tinikul R, Chaiyen P (2014) Coordination of reduced flavin transfer between the proteins of two-component flavin-dependent monooxygenases. Arch Biochem Biophys 555–556:33–46
65. Suzuki K, Kaidoh T, Katagiri M, Tsuchiya T (1983) O_2 incorporation into a long-chain fatty acid during bacterial luminescence. Biochim Biophys Acta 722:279–301
66. Szittner R, Meighen E (1990) Nucleotide sequence, expression, and properties of luciferase coded by *lux* genes from a terrestrial bacterium. J Biol Chem 265:16581–16587
67. Tinikul R, Pitsawong W, Sucharitakul J, Nijvipakul S, Ballou DP, Chaiyen P (2013) The transfer of reduced flavin mononucleotide from LuxG oxidoreductase to luciferase occurs via free diffusion. Biochemistry 52:6834–6843
68. Tinikul R, Thotsaporn K, Thaveekarn W, Jitrapakdee S, Chaiyen P (2012) The fusion *Vibrio campbellii* luciferase as a eukaryotic gene reporter. J Biotechnol 162:346–353
69. Tongsook C, Sucharitakul J, Thotsaporn K, Chaiyen P (2011) Interactions with the substrate phenolic group are essential for hydroxylation by the oxygenase component of *p*-hydroxyphenylacetate 3-hydroxylase. J Biol Chem 286:44491–44502
70. Tu S-C (1982) Isolation and properties of bacterial luciferase intermediates containing different oxygenated flavins. J Biol Chem 257:3719–3725
71. Ulitzur S, Hastings JW (1979) Evidence for tetradecanal as the natural aldehyde in bacterial bioluminescence. Proc Natl Acad Sci USA 76:265–267

72. Valkova N, Szittner R, Meighen EA (1999) Control of luminescence decay and flavin binding by the LuxA carboxyl-terminal regions in chimeric bacterial luciferases. Biochemistry 38:13820–13828
73. van Berkel WJH, Kamerbeek NM, Fraaije MW (2006) Flavoprotein monooxygenases, a diverse class of oxidative biocatalysts. J Biotechnol 124:670–689
74. Villa R, Willetts A (1997) Oxidations by microbial NADH plus FMN-dependent luciferases from *Photobacterium phosphoreum* and *Vibrio fischeri*. J Mol Catal B: Enzym 2:193–197
75. Werlund-Karlsson A, Saviranta P, Karp M (2002) Generation of thermostable monomeric luciferases from *Photorhabdus luminescens*. Biochem Biophys Res Commun 296:1072–1076
76. Xi L, Cho K-W, Herndon ME, Tu S-C (1990) Elicitation of an oxidase activity in bacterial luciferase by site-directed mutation of a noncatalytic residue. J Biol Chem 265:4200–4203
77. Xin X, Xi L, Tu S-C (1991) Functional consequences of site-directed mutation of conserved histidyl residues of the bacterial luciferase α subunit. Biochemistry 30:11255–11262
78. Xin X, Xi L, Tu S-C (1994) Probing the *Vibrio harveyi* luciferase β subunit functionality and the intersubunit domain by site-directed mutagenesis. Biochemistry 33:12194–12201

Part II
Applications of Bioluminescence in Environment and Security

Adv Biochem Eng Biotechnol (2016) 154: 77–100
DOI 10.1007/10_2015_332

Published Online: 17 October 2015

Detection of Metal and Organometallic Compounds with Bioluminescent Bacterial Bioassays

M.J. Durand, A. Hua, S. Jouanneau, M. Cregut and G. Thouand

Abstract Chemical detection of metal and organometallic compounds is very specific and sensitive, but these techniques are time consuming and expensive. Although these techniques provide information about the concentrations of compounds, they fail to inform us about the toxicity of a sample. Because the toxic effects of metals and organometallic compounds are influenced by a multitude of environmental factors, such as pH, the presence of chelating agents, speciation, and organic matter, bioassays have been developed for ecotoxicological studies. Among these bioassays, recombinant luminescent bacteria have been developed over the past 20 years, and many of them are specific for the detection of metals and metalloids. These bioassays are simple to use, are inexpensive, and provide information on the bioavailable fraction of metals and organometals. Thus, they are an essential complementary tool for providing information beyond chemical analysis. In this chapter, we propose to investigate the detection of metals and organometallic compounds with bioluminescent bacterial bioassays and the applications of these bioassays to environmental samples.

M.J. Durand (✉) · A. Hua · S. Jouanneau · G. Thouand
University of Nantes, UMR CNRS GEPEA 6144, 18 Bd Gaston Defferre,
85000 La Roche sur Yon, France
e-mail: marie-jose.durand-thouand@univ-nantes.fr

M. Cregut
Capacités SAS, 26 Bd Vincent Gâche, 44200 Nantes, France

Graphical Abstract

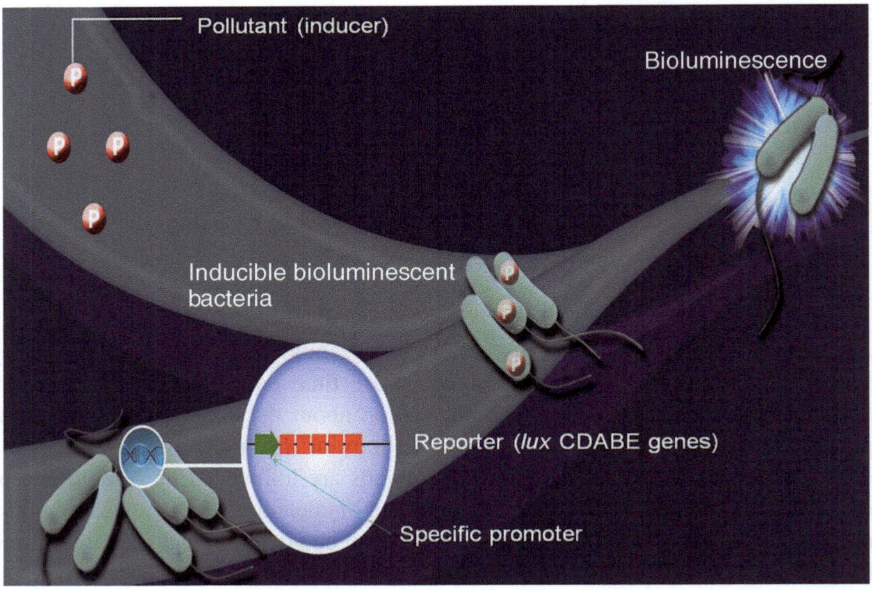

Keywords Bioluminescent bacterial bioassay · Metals and organometallics · Specific detection

Contents

1 Introduction

Many aspects of microbial biosensing and reporter gene technology have been intensively reviewed over the last 10 years [12, 18, 22, 39, 54, 55, 63, 65, 66, 69, 70]. Contamination of metals and organometallic compounds in the environment is a major global concern because they have the potential to affect human health and ecosystems. Due to their toxicity, government agencies (e.g., the EPA, WHO, EU) regulate the concentrations of these compounds in groundwater, drinking water, foods, and the like. To comply with these regulations, detection of metals is accomplished by chemical methods such as atomic absorption, furnace atomic absorption, and atomic emission. These methods are very sensitive but are time consuming and require a pretreatment of the sample. To cope with the need for fast and simple methods, bioassays were developed; among them, a large proportion involved bioluminescent bacteria.

2 Metals, Metalloids, and Organometallic Compounds

2.1 Definition

Metals are naturally present in the environment. Nevertheless, anthropogenic activity may increase their concentration, leading to pollution of the ecosystem. This pollution is due to many sources, such as urbanisation and industrial activities, especially in developing countries [13, 34, 50]. Pollution may also be due to contamination from treatment plant wastewaters or mining activities [2]. Many toxic metals have been referred to as heavy metals, but the definition remains unclear and without scientific reason or legal enforcement. Commonly, definitions of heavy metals describe them as elements with densities higher than 5000 kg/m^3 and as being associated with environmental effects and potential toxicity or ecotoxicity. This denomination includes not only ‘real heavy metals’, such as mercury (Hg) or lead (Pb), but also other elements that are not particularly ‘heavy’, such as zinc (Zn) and aluminium (Al). The heavy metal denomination also includes metalloids [1]. Metalloids are semi-metallic elements whose physical and chemical characteristics are intermediate between metals and nonmetals, and most of them are semiconductors. In this category, we find arsenic (As), antimony (Sb), tellurium (Te), selenium (Se), and polonium (Po).

To avoid confusion regarding the significance of the term ‘heavy metal’, the metallic trace element (MTE) denomination is progressively substituted for the heavy metal notion.

Organometallic compounds are defined as compounds containing a covalent bond between a carbon atom and a metal. They exhibit properties that are common to both organic substances and metals. The origin of these compounds may be anthropogenic, for example, organotin (tributyltin used as an antifouling agent), or environmental, for example, biomethylation of mercury by microorganisms. Generally, methylation of metals increases their toxicity, except for arsenic and selenium.

2.2 Classification and Toxicity

Among MTE, it is necessary to separate metals and metalloids that are essential to biological life and act in many metabolic pathways (e.g., copper, zinc, selenium) from the nonessential MTE without any biological function. These two groups are shown on the periodic table of elements (Fig. 1). Nevertheless, all MTE have toxicity with respect to living organisms according to their concentrations.

Because of their toxic properties, carcinogenic, neurotoxic, or endocrine disrupting, metallic (As, Pb, Cd, Ni, Hg, Al) and organometallic (organomercury and organotin) substances are considered priority substances to be detected in water by the European Union (2013/39/EU directive). The World Health Organization has also defined guideline values for MTE in drinking water (see Table 1) [68].

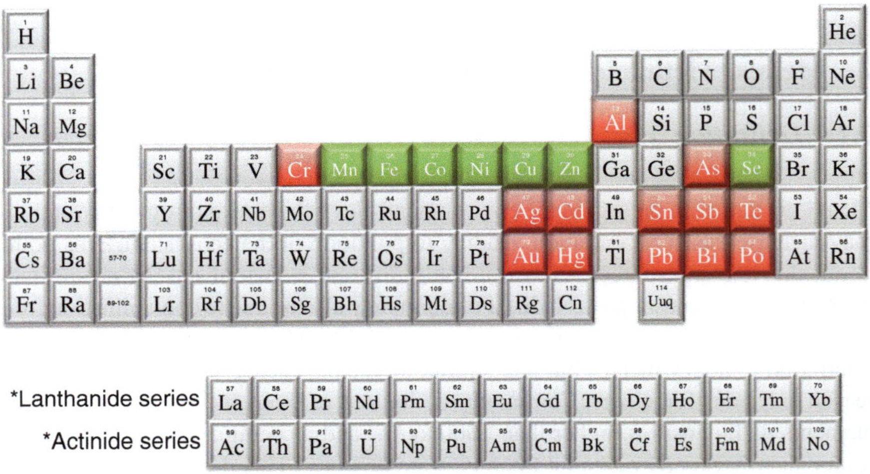

Fig. 1 A periodic table of MTE. *Green* indicates essential metal (loids); *red* indicates MTE that are toxic or have no benefits in most organisms

Table 1 Guideline values for MTE in drinking water according to the WHO

Compound	Guideline Value
Antimony (Sb)	20 μg L^{-1}
Arsenic (As)	10 μg L^{-1}
Cadmium (Cd)	3 μg L^{-1}
Chromium (Cr)	50 μg L^{-1}
Copper (Cu)	2 mg L^{-1}
Lead (Pb)	10 μg L^{-1}
Mercury, inorganic (Hg)	6 μg L^{-1}
Nickel (Ni)	70 μg L^{-1}

3 Bioluminescent Bacteria for the Detection of Specific Compounds

3.1 Bacterial Bioluminescence

Many bacteria sensitive to specific metals and/or metalloids have been constructed within the last 20 years. All the bioluminescent bioreporters described in the literature carried out fusion between reporter genes (*luxAB* or *luxCDABE*; a few of them harbour a *luc* fusion but are not included in this chapter) and the promoter of genes involved in metal resistance mechanisms.

Bioluminescence is an enzymatic production of light by living organisms [16, 40]. Bacterial luminescence has been found to occur naturally in 11 bacterial species from five genera: *Vibrio*, *Aliivibrio*, *Photobacterium*, *Shewanella*, and *Photorhabdus* [16, 64]. The reaction is controlled by five gene operon denoted as *luxCDABE*. *luxA* and *luxB* are encoded for the luciferase, and the three other genes *luxCDE* are involved in the production and turnover of the aldehyde. The *luxAB* bioreporter requires the addition of aldehyde (n-decanal) to produce bioluminescence, whereas the *luxCDABE* bacteria can generate luminescence spontaneously.

The molecular biology of these genes and enzymes has been well studied in *Vibrio harveyi*, *Aliivibrio fischeri*, *Photobacterium phosphoreum*, and *Photobacterium leiognathi* [42, 43]. Because luminescence is a sensitive reaction easily measurable by commercial devices such as the CCD camera or luminometer, the *luxAB* or *luxCDABE* genes are widely used as reporter genes to construct bioreporters. Baldwin et al. [3] were the first to use the *luxAB* genes of *V. harveyi* in *Escherichia coli* as a reporter of toxicant presence. From these results, the *luxAB* or *luxCDABE* reporter genes were applied to create various bioreporter bacteria to detect specific compounds such as metals, organic substances, or general metabolic stresses.

3.2 Strategy for the Construction of Recombinant Bioluminescent Bacteria

Two main strategies can be used to create inducible bioluminescent bacteria. First, bacterial genes involved in the resistance to metals are known. The *lux* genes, without promoter, are inserted downstream of the promoter of the known gene. When the compound is present, the promoter of the resistance gene is induced, and the luciferase is produced concomitantly. Luciferase expression is under the control of the resistance gene, and its expression is generally proportional to the concentration of the bioaccessible compounds (Fig. 2, left diagram).

Second, luciferase gene insertion can be applied to find new, uncharacterised genes [25, 26]. In this case, the luciferase reporter gene is randomly inserted into a host bacterium (*E. coli*, e.g.) to ensure only one insertion into the bacterial

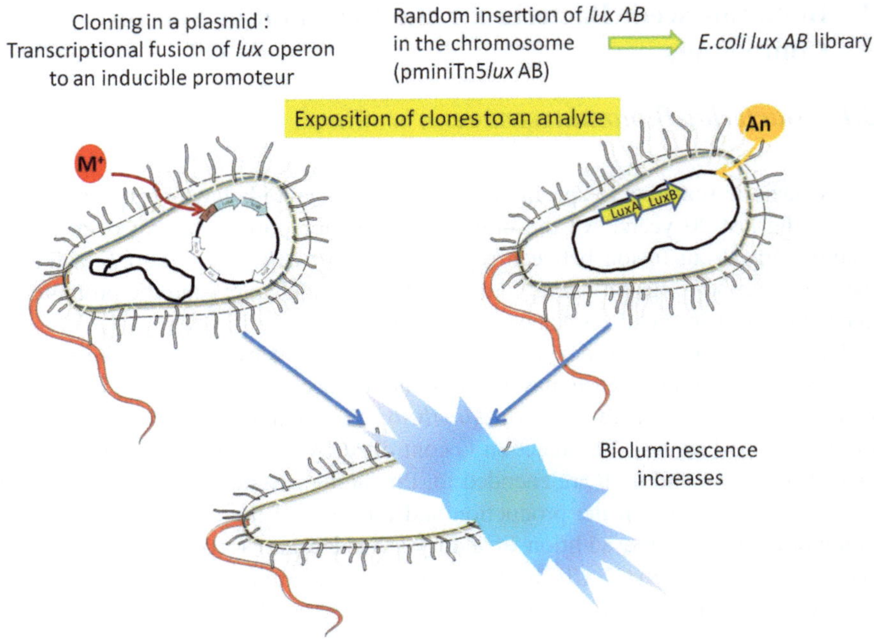

Fig. 2 Two main strategies for designing specific bioluminescent bacteria: (i) in the *left* diagram, the *lux* operon (in *blue*) without is promoter is inserted downstream of the promoter of the known gene (in *red*), or (ii) in the *right* diagram, the luciferase genes (*luxAB*) are randomly inserted into the bacterial chromosome. In the presence of the analyte, bioluminescence increases and allows its quantification (decanal long chain aldehyde is only necessary when using *luxAB* genes)

chromosome (Fig. 2, right diagram). The resultant clones are stored in a bank ('library'), and each clone can then be tested against a particular compound [26, 59].

4 Metal Bioreporters

As mentioned above, the construction of bioreporters for metal detection requires knowledge of the MTE regulation by the cells.

4.1 Bacterial Resistance Mechanisms

Metals are ubiquitous in the environment, but environmental factors can greatly influence their concentrations. To survive, bacteria must monitor their environment to control the import of essential MTE and to reduce the uptake or remove non-essential MTE. Homeostasis of essential MTE concentrations is mediated by

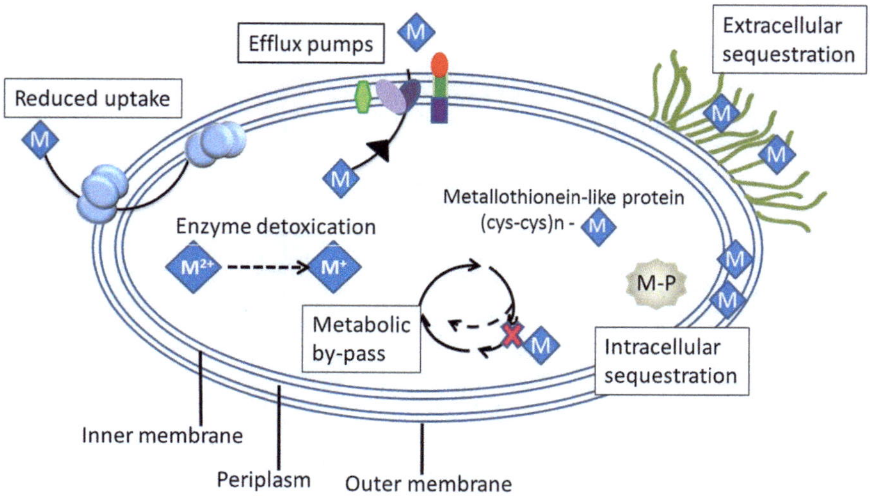

Fig. 3 Mechanisms of adaptation and resistance of bacteria to nonessential MTE (M). Bacteria could decrease the expression and activity of influx pumps (reduced uptake) and expressed efflux proteins (efflux pumps). Some microorganisms upregulate the expression of extracellular polymers to trap metals (extracellular sequestration). Bacteria could precipitate metals as metal salts, these particulates could be associated with the cytoplasmic membrane, and MTE could be bound by cytoplasmic proteins as, for example, metallothioneins (intracellular sequestration). Cells could also produce alternative enzymes with catalytic cores that do not bind to toxic MTE (metabolic bypass). Metals can undergo specific enzymatic reactions to produce fewer toxic species or less available forms (enzyme detoxication; adapted from Maier et al. [40], Lemire et al. [28])

proteins involved in metal uptake (influx pump) and storage processes. Most of these systems are constitutive (chromosomally encoded) and could be repressed to decrease the uptake when the cytosolic concentration of a metal becomes too high [15, 47, 57].

Because metals have similar properties, nonessential MTE could be transported by the influx pump. To maintain these metals under toxic concentrations, bacteria have developed mechanisms of resistance and adaptation. These mechanisms are diverse in terms of sequestration, efflux pumps, or chemical modification (Fig. 3).

Among these mechanisms, efflux pumps are effective at removing this excess of nonessential MTE. These membrane transporters could be encoded chromosomally or by mobile genetic elements. The expression of these systems is controlled by regulators that bind metals with high affinity. The activity of the efflux pump may be due to ATP hydrolysis or chemiosmotic potential. Figure 4 shows the major system of proteins involved in MTE resistance in Gram-negative bacteria.

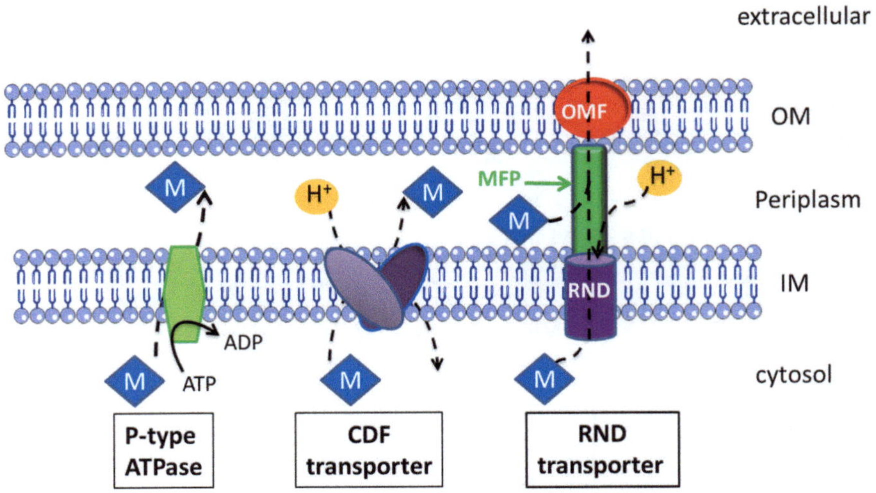

Fig. 4 Mechanism of efflux found in Gram-negative bacteria. outer membrane (*OM*); inner membrane (*IM*). The P-type ATPase superfamily transports ions against the concentration gradient. P-type ATPases are found in all life kingdoms. P_{1B}-ATPases are a subgroup of P-ATPases that transport MTE (Cu^{+}, Cu^{2+}, Ag^{+}, Zn^{2+}, Cd^{2+}, Pb^{2+}, Co^{2+}). They are found in the inner membrane and use ATP energy to pump metal ions from the cytoplasm to the periplasm. CDF family transporters (cation diffusion facilitator) act as chemiosmotic ion protons and are involved in the influx or efflux of MTE. Resistance-nodulation-cell-division (*RND*) efflux systems help Gram-negative bacteria to resist high metal concentrations. These efflux systems are formed by a central RND proton-substrate antiporter, a membrane fusion protein (*MFP*), and an outer membrane factor (*OMF*). These protein complexes span the complete cell wall and could catalyse transenvelope efflux or efflux from the periplasm to the outside [36–38]

4.2 Bioluminescent Bioreporters

Currently, more than 28 bioluminescent bioreporters have been described in the literature for sensing metals based on the transcriptional fusion of *lux* genes with the characterised genes involved in bacterial metal resistance. The fusion is integrated into the chromosome of the host strain or, in most cases, into a medium or high-copy–number plasmid.

As mentioned in Fig. 2, some bioreporters have been obtained by random insertion of *luxAB* into the chromosome. Guzzo et al. [25] reported the detection of aluminium in water by an *E. coli* containing *fliC*::Tn5-*luxAB* gene fusion. The most used host strain is the well-known bacteria *E. coli*, but various other species can integrate genomic fragments leading to metal resistance, such as the *Pseudomonas*, *Staphylococcus*, and *Bacillus* genera, or, more recently, *Ochrobactrum tritici*, a soil isolated bacteria [7].

Table 2 reported the strains found in the literature. The host strain, the origin of the promoter, the analyte detected, and a brief description of the bioassay conditions are reported. The *luxCDABE* operon from *Photorhabdus luminescens* is the most

Table 2 Construction of lux bioreporters for metal detection, and their applications

Promoter (origin)	Reporter gene (origin)	Location	Chemical target analyte	Detection range, linear response	Cultivation methods	Conditions of induction	Application to environmental samples	Reference
Host strain *Escherichia coli*								
arsR (pBGD23)	*luxAB* (*Vibrio harveyi*)	pJAMA	As^{3+}	8–80 μg L^{-1}	LB medium, 37 °C, OD_{600nm} = 0.5–0.6	90 min, 30 °C, n-decanal	Tap water	[56]
arsR (pR773)	*luxCDABE* (*Photorhabdus luminescens*)	pTOO31	As^{3+}	8 μg L^{-1}	M9 medium, 37 °C, OD_{600nm} = 0.5	3 h	Soil and sediments	[30]
			As^{5+}	141 μg L^{-1}				
			As^{3+}	18 μg L^{-1}		Immobilised on optical fibres		
			As^{5+}	141 μg L^{-1}				
arsR (*E. coli*)	*luxCDABE* (*Aliivibrio fischeri*)	pBtac	As^{3+}	0.4 μg L^{-1}	Acetate medium, 30 °C, OD_{600nm} = 0.5		Tap water	[11]
			As^{5+}	4 μg L^{-1}				
copA (*E. coli*)	*luxCDABE* (*A. fischeri*)	pUCD615	Cu^{2+}	19–63 μg L^{-1}	LB medium, 30 °C, OD_{600nm} = 10, then dilution in GGM medium, OD_{600nm} = 0.2	80 min, 30 °C	–	[52]
			Ag^{+}	10 μg L^{-1}				
cueR and *copA* (*E. coli*)	*luxCDABE* (*P. luminescens*)	pDN18N	Cu^{2+}	0.02 mg L^{-1}	M9 medium, 30 °C, shaking 200 rpm, OD_{600nm} = 0.6–0.8, then dilution OD_{600nm} = 0.1	120 min, 30 °C	–	[31]
			Ag^{+}	0.3 mg L^{-1}				
merR (Tn21)	*luxCDABE* (*P. luminescens*)	PTOO11	Hg^{2+}	0.003 μg L^{-1}	M9 medium, 30 °C, shaking 200 rpm, OD_{600nm} = 0.5	3 h	Soil and sediments	[30]
				2.6 μg L^{-1}		Immobilised on optical fibres		
merR (plasmid)	*luxCDABE* (*P. luminescens*)	pSL1190	Hg^{2+}	0.03 μg L^{-1}	M9 medium, 30 °C, shaking 200 rpm, OD_{600nm} = 0.6–0.8, then dilution OD_{600nm} = 0.1	120 min, 30 °C	–	[31]
			Methylmercury	2 ng L^{-1}				
			Cd^{2+}	0.06 mg L^{-1}				

(continued)

Table 2 (continued)

Promoter (origin)	Reporter gene (origin)	Location	Chemical target analyte	Detection range, linear response	Cultivation methods	Conditions of induction	Application to environmental samples	Reference
zntA (*E. coli*)	*luxCDABE* (*A. fischeri*)	pUCD615	Cd^{2+}	Up to 33.8 μg L^{-1}	LB medium, 30 °C, OD_{600nm} = 10, then dilution in GGM medium, OD_{600nm} = 0.2	80 min, 30 °C	–	[52]
			Pb^{2+}	Up to 207 μg L^{-1}				
			Zn^{2+}	Up to 2 mg L^{-1}				
			Hg^{2+}	Up to 6 mg L^{-1}				
zntR and *zntA* (*E. coli*)	*luxCDABE* (*P. luminescens*)	pDN18N	Hg^{2+}	0.02 mg L^{-1}	M9 medium, 30 °C, shaking 200 rpm, OD_{600nm} = 0.6–0.8, then dilution OD_{600nm} = 0.1	120 min, 30 °C	–	[31]
			Cd^{2+}	0.003 mg L^{-1}				
			Zn^{2+}	0.8 mg.L^{-1}				
			Pb^{2+}	0.2 mg.L^{-1}	HMM medium, 30°C, shaking 200 rpm,			
Host strain *Pseudomonas fluorescens*								
merR (plasmid)	*luxCDABE* (*P. luminescens*)	pDN18 N	Hg^{2+}	0.3 μg L^{-1}	M9 medium, 30 °C, shaking 200 rpm, OD_{600nm} = 0.6–0.8, then dilution OD_{600nm} = 0.1	120 min, 30 °C 120 min, 30 °C	–	[31]
			Methylmercury	0.03 μg L^{-1}				
			Cd^{2+}	0.9 mg L^{-1}				
		chr	Hg^{2+}	0.8 μg L^{-1}	M9 medium, 30 °C, shaking 200 rpm, OD_{600nm} = 0.4			
			Methylmercury	0.8 μg L^{-1}				
			Cd^{2+}	9.2 mg L^{-1}				
cueR and *copA* (*E. coli*)	*luxCDABE* (*P. luminescens*)	chr	Cu^{2+}	8 mg L^{-1}	M9 medium, 30 °C, shaking 200 rpm, OD_{600nm} = 0.8	120 min, 30 °C	–	[31]
zntR and *zntA* (*E. coli*)	*luxCDABE* (*P. luminescens*)	chr	Hg^{2+}	8 μg L^{-1}	M9 medium, 30 °C, shaking 200 rpm, OD_{600nm} = 0.2	120 min, 30 °C	–	[31]
			Cd^{2+}	0.02 mg L^{-1}				
			Zn^{2+}	3.2 mg L^{-1}				
			Pb^{2+}	0.07 mg L^{-1}	HMM medium, 30 °C, shaking 200 rpm			

(continued)

Table 2 (continued)

Promoter (origin)	Reporter gene (origin)	Location	Chemical target analyte	Detection range, linear response	Cultivation methods	Conditions of induction	Application to environmental samples	Reference
cadRA (*Pseudomonas putida*)	*luxCDABE* (*P. luminescens*)	pDN18N	Hg^{2+}	0.03 mg L^{-1}	M9 medium, 30 °C, shaking 200 rpm, OD_{600nm} = 0.6–0.8, then dilution OD_{600nm} = 0.1	120 min, 30 °C	–	[31]
			Cd^{2+}	00.01 mg L^{-1}				
			Zn^{2+}	1.3 mg L^{-1}				
			Pb^{2+}	0.1 mg L^{-1}	HMM medium, 30 °C, shaking 200 rpm			
		chr	Hg^{2+}	0.005 mg L^{-1}	M9 medium, 30 °C, shaking 200 rpm OD_{600nm} = 0.8			
			Cd^{2+}	0.006 mg L^{-1}				
			Zn^{2+}	0.6 mg L^{-1}				
			Pb^{2+}	0.1 mg L^{-1}	HMM medium, 30 °C, shaking 200 rpm			
pbrRA (*Cupriavidus metallidurans CH34*)	*luxCDABE* (*P. luminescens*)	pDN18N	Hg^{2+}	0.04 mg L^{-1}	M9 medium, 30 °C, shaking 200 rpm, OD_{600nm} = 0.6–0.8, then dilution OD_{600nm} = 0.1	120 min, 30 °C	–	[31]
			Cd^{2+}	0.2 mg L^{-1}				
			Zn^{2+}	6.4 mg L^{-1}				
			Pb^{2+}	0.3 mg L^{-1}	HMM medium, 30 °C, shaking 200 rpm			
		chr	Hg^{2+}	0.008 mg L^{-1}	M9 medium, 30 °C, shaking 200 rpm, OD_{600nm} = 0.6–0.8, then dilution OD_{600nm} = 0.1			
			Cd^{2+}	0.02 mg L^{-1}				
			Zn^{2+}	3.2 mg.L^{-1}				
			Pb^{2+}	0.07 mg.L^{-1}	HMM medium, 30°C, shaking 200 rpm			
zntR and *zntA* (*E. coli*)	*luxCDABE* (*P. luminescens*)	chr	Hg^{2+}	8 µg L^{-1}	M9 medium, 30 °C, shaking 200 rpm, OD_{600nm} = 0.2	120 min, 30 °C	–	[31]
			Cd^{2+}	0.02 mg L^{-1}				
			Zn^{2+}	3.2 mg L^{-1}				
			Pb^{2+}	0.07 mg L^{-1}	HMM medium, 30 °C, shaking 200 rpm			

(continued)

Table 2 (continued)

Promoter (origin)	Reporter gene (origin)	Location	Chemical target analyte	Detection range, linear response	Cultivation methods	Conditions of induction	Application to environmental samples	Reference
Host strain *Pseudomonas putida*								
cadA1 (*P. putida*)	*luxCDABE* (*P. luminescens*)	pDN18N	Zn^{2+}	58 μg L^{-1}	HMM medium, 30 °C, 3 h, shaking, early exponential phase	60 min, 30 °C	Soil	[29]
czcCBA (*P. putida*)	*luxCDABE* (*P. luminescens*)	pDN18N	Zn^{2+}	10 μg L^{-1}	HMM medium, early exponential phase	60 min, 30 °C	Soil	[29]
			Pb^{2+}	187 μg L^{-1}				
			Cd^{2+}	125 μg L^{-1}				
pupA (*P. putida*)	*luxCDABE* (*A. fischeri*)	chr	Fe^{3+}	0.55–55 μg L^{-1}	KB medium, 30 °C, OD_{600nm} = 0.8	30 °C	–	[35]
Host strain *Staphylococcus aureus*								
cadCA (pI258)	*luxCDABE* (*P. luminescens*)	pTOO24	Hg^{2+}	0.003 mg L^{-1}	LB medium, 30 °C, OD_{600nm} = 0.8–1, then dilution M9 medium OD_{600nm} = 0.1	120 min, 30 °C	–	[31]
			Cd^{2+}	0.007 mg L^{-1}				
			Zn^{2+}	3.2 mg L^{-1}				
			Pb^{2+}	0.03 mg L^{-1}	LB medium, 30 °C, OD_{600nm} = 0.8–1, then dilution HMM medium OD_{600nm} = 0.1			
Host strain *Bacillus subtilis*								
cadCA (pI258)	*luxCDABE* (*P. luminescens*)	pTOO24	Hg^{2+}	0.01 mg L^{-1}	LB medium, 30 °C, OD_{600nm} = 0.8–1, then dilution M9 medium OD_{600nm} = 0.1	120 min, 30 °C	–	[31]
			Cd^{2+}	0.002 mg L^{-1}				
			Zn^{2+}	1.6 mg L^{-1}				
			Pb^{2+}	0.03 mg.L^{-1}	LB medium, 30 °C, OD_{600nm} = 0.8–1, then dilution HMM medium OD_{600nm} = 0.1			

(continued)

Table 2 (continued)

Promoter (origin)	Reporter gene (origin)	Location	Chemical target analyte	Detection range, linear response	Cultivation methods	Conditions of induction	Application to environmental samples	Reference
Host strain *Synechococcus* sp. PCC 7002								
isiAB (*Synechococcus* sp. PCC 7002)	*luxAB* (*V. harveyi*)	pAM1414	Iron	–	Medium A, exponential phase, light (45 µmol m^{-2} s^{-1})	12 h, 25 °C, n-decanal	Water	[6]
Host strain *Synechococcus* sp. PCC 7942								
smtA (*Synechococcus* sp. PCC 7942)	*luxCDABE* (*A. fischeri*)	pJLE23	Zn^{2+}	32.5–260 µg L^{-1}	BG-11 medium, 30 °C, OD_{730nm} = 0.15, early exponential phase, light	60 min, room temperature	–	[19]
Host strain *Synechococcus* sp. PCC 6803								
coaT (*Synechocystis* sp. PCC 6803)	*luxAB* (*V. harveyi*)	pND6	Co^{2+}	18 µg L^{-1}–1.5 mg L^{-1}	BG-11 medium, 30 °C, OD_{720nm} = 0.6, early exponential phase, light (40 µmol m^{-2} s^{-1})	3 h, 25 °C, darkness, in a 3 % CO_2 enriched atmosphere	Soil	[48]
			Zn^{2+}	13 µg L^{-1}–845 µg L^{-1}				
nrsBACD (*Synechocystis* sp. PCC 6803)	*luxAB* (*V. harveyi*)	pND6	Ni^{2+}	12–350 µg L^{-1}	BG-11 medium, 30 °C, mid exponential phase, OD_{720nm} = 0.6, early exponential phase, light (40 µmol m^{-2} s^{-1})	3 h, 25 °C, light (40 µmol m^{-2} s^{-1})	Soil	[48]
Host strain *Cupriavidus metallidurans* CH34								
cnrYXH (*C. metallidurans CH34*)	*luxCDABE* (*A. fischeri*)	pMOL877	Ni^{2+}	6 µg L^{-1}–3.5 mg L^{-1}	–	23 °C	Soil	[60, 61]
			Co^{2+}	0.5–23 mg L^{-1}				

Chr: Single chromosomal insertion

used transcriptional reporter, which is probably due to the higher stability of luciferase, as mentioned by Meighen [42]. As we can see in this table, there are different media that could be used (rich medium such as LB or minimum medium). The numbers of cells exposed are also very different according to the authors; due to this difference, it is difficult to compare the construction. Nevertheless, for most of the bioreporters, responses were dose-dependent with a linear response after an incubation period with the sample of 60–180 min. The highest sensitivity found concerns the detection of mercury by *E. coli* pDL20 [49], where the detection limit is approximately 0.05 nM of Hg^{2+} (10 ng L^{-1}). Most of the applications concern a standard solution in water or environmental samples supplemented with a standard solution of metals.

5 Development of Bioassay

5.1 Experimental Procedures

The detection of compounds by specific bioreporters can be affected by inducible resistance systems, experimental methods, sample preparation, and the composition of the sample. To increase the sensitivity of detection and to minimise the variability of the response, the assay procedure must be controlled.

To obtain detection limits for metals close to the guidelines of the agency (the EPA, WHO, EU), resistance systems coming from less resistant bacteria, such as *E. coli*, should be preferred. For example, the strain *Cupriavidus metallidurans* CH34 (*Ralstonia metallidurans* CH34 [44]), which harbours many resistance systems from pMOL28 and pMOL30 plasmids, has been modified by the insertion of the transposon Tn4431 containing the *luxAB* genes [14, 58]. The detection limit of the strain obtained was 1 mM for Pb^{2+} compared with 30 nM Pb^{2+} for other engineered bacteria such as *E. coli* and *Staphylococcus aureus*, which are naturally less resistant to metals [39]. Another strategy to increase the sensitivity of the detection consists of an inactivation of genes involved in the detoxification mechanism. For example, Pepi et al. [49] compared the detection limit of two different *mer*::*lux* fusions. The first one, described by Selifonova et al. [53], contained the whole *mer* operon (six functional genes *merRTPCAD*), where *merA* encodes the mercuric reductase (MR) enzyme, which is responsible for the reduction of Hg^{2+} to the less toxic Hg°. The second construction harbours the *mer* operon regulatory element, but with a partial deletion of *merA* leading to a non-functional mercury reductase (MR). The second bioreporter was found to be 27 times more sensitive than the first.

Experimental procedures could also affect the sensitivity of the bioassay. Indeed, cells must be cultivated in controlled media, and minimum media are generally used to increase the limit of detection [17]. Because luminescence could also be dependent on the growth phase of the cell [11], induction of luminescence must be recorded during the growth, generally early or middle growth phase, to allow the best signal to be obtained. Table 3 highlights the diversity of experimental procedures for the detection of arsenic with engineered *E. coli.*

In the literature, most of the bioassays have been conducted according to batch culture cells or from freeze-dried cells. Freeze-drying technology offers convenient long-term storage (several months at −20 °C, with reproducible detection results) and provides the opportunity to have a bioassay ready to use for an end-user [5, 67].

Ren and Frymier [51] emphasised that variability of the result with bioluminescent bacteria could be avoided by a clear definition of the protocol. They noted that the storage conditions of the cells (−80 °C) contribute to a major part of the variability of the bioassay. For long-duration storage, freeze-drying treatment did not modify the sensitivity of the bioassay [17, 58]. Figure 5 shows the major factors that must be controlled in the assay protocol to reduce the variability of the response.

5.2 Application to Environmental Samples

Even if legislation requires the accurate concentration of total metal, there is still a need to determine the toxic effect of the compound. Total metal concentration is not necessarily related to the metal's bioaccessibility, biodisponibility, and toxicity. Furthermore, the toxicity of a metal is highly related to its chemical form; for example, cations can accumulate rapidly in certain organisms, whereas a metal bound to particles could be less toxic. To check the bioavailability of the metal, bioassays using genetically engineered bacteria have been used for the last 20 years. Applications have been used mainly for water contamination, and some applications have been done on soil (Table 2) and food (determination of arsenic in rice [4]).

All studies show that bioluminescent bioreporters are not specific to a metal, but several metals could induce luminescence, depending on the sensing element. The detection limit and specificity of four bioreporters have been characterised under the same conditions; the results show that none of the bacteria are specific to one metal (Table 4). Strain detection limits and specificities vary from tested compounds, but for three main metalloid compounds (cadmium, mercury, and copper), there is at least one bacterial strain that allows detection below ES: *E. coli* Zntlux for cadmium detection, Merlux for both mercury and copper detection, and Coplux for copper detection. Note that even if the Arslux bacterial strain does not detect As(III) and As (V) with great sensitivity (detection 18 times higher than the permissive levels), this strain is specific to arsenic detection at high concentrations and could be an alternative for detecting arsenic in the water of highly contaminated geographical areas such as Bangladesh [45].

As seen above, a given bacterial strain can respond to many compounds, and conversely, a given compound can be detected by several reporter bacteria [31]. These cross-detections lead to difficulties in signal analysis and compound identification. However, as reported by Jouanneau et al. [32] it is possible to improve

Table 3 Detection of arsenic by different bioelements in the *E. coli* host strain

Promoter (origin)	Reporter gene (origin)	Chemical target analyte	Detection range, linear response	Cultivation methods	Conditions of induction	Reference
arsB (pI258)	*luxCDABE* (*A. fischeri*)	As^{3+}	–	LB medium, 37 °C, mid exponential	60 min, 30 °C, orbital shaker 190 rpm, aldehyde	[20]
		As^{5+}	–			
arsR (pR773)	*luxCDABE* (*P. luminescens*)	As^{3+}	18 μg L^{-1}	M9 medium, 37 °C, OD_{600nm} = 0.5	3 h	[30]
		As^{5+}	141 μg L^{-1}			
arsB (*E. coli*)	*luxAB* (*V. harveyi*)	As^{5+}	0.1 mg L^{-1}	LB medium, 37 °C, OD_{600nm} = 0.3–0.4	120 min	[10]
arsR (pBGD23)	*luxAB* (*V. harveyi*)	As^{3+}	8–80 μg L^{-1}	LB medium, 37 °C, OD_{600nm} = 0.5–0.6	90 min, 30 °C, 190 rpm, n-decanal (2 mM)	[56]
arsR (pBGD23)	*luxAB* (*V. harveyi*)	As^{3+}	10–100 μg L^{-1}	LB medium, 37 °C, OD_{600nm} = 0.5	90 min, 30°C, 200 rpm, n-decanal (18 mM)	[62]
arsR (pBGD23)	*luxAB* (*V. harveyi*)	As^{3+}	0.02–0.15 μg equivalent per gram of dry weight (rice)	LB medium, 37 °C, OD_{600nm} = 0.6	60 min, 30 °C, 500 rpm, n-decanal (2 mM)	[4]
arsR (plasmid)	*luxCDABE* (*P. luminescens*)	As^{3+}	8 μg L^{-1}	M9 medium, 37 °C, OD_{600nm} = 0.5	2 h, 37 °C, without shaking	[27]
		As^{5+}	80 μg L^{-1}			
arsR (*E. coli*)	*luxCDABE* (*A. fischeri*)	As^{3+}	0.4 μg L^{-1}	Acetate medium, 30 ° C	Immobilisation in agarose matrix	[11]
		As^{5+}	4 μg L^{-1}			

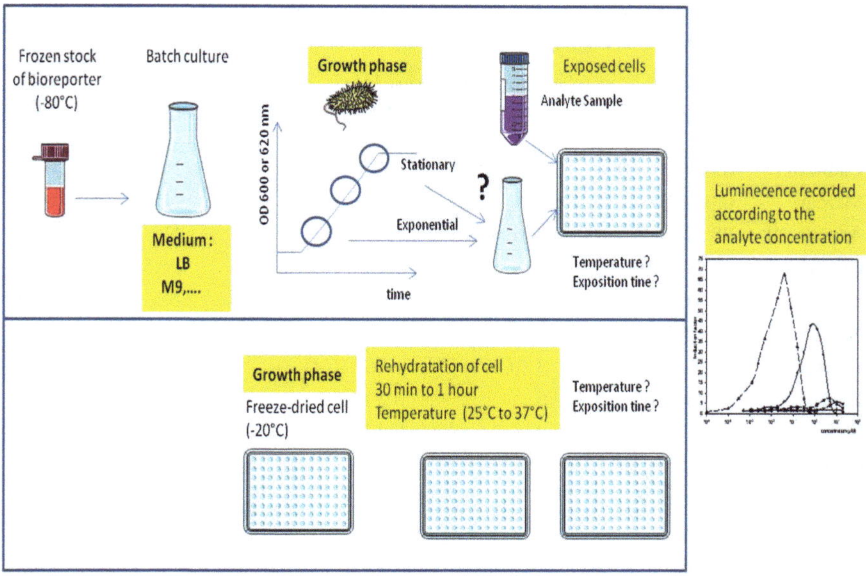

Fig. 5 Bioassay protocol for the detection of metal or organometallic compounds by specific bioluminescent bacteria. After batch culture, cells could be freeze-dried in a microplate in an adequate medium with cryoprotectant, where storage is generally at −20°C. After rehydration with water, the cells are ready to be exposed to the sample. For the fresh cells, luminescence is recorded after an exposition time, generally from 30 to 180 min. In addition to batch culture cells and freeze-dried cells, some authors have developed an alternative method for bacterial cell trapping, such as agarose or alginate-bead immobilised bacterial or sol-gel entrapment. These methods are generally developed when bioelements are used in a biosensor device

sensibility, specificity, and compound identification by a statistical approach coupled with a set of reporter bacteria, as described in Table 4. Crossings between different detection ranges with a set of bioluminescent bacteria, in addition to statistical analysis, allowed the identification and quantification of metallic compounds with a high correlation: 98 and 94 %, respectively. This set of bacteria was applied on leachate from real polluted soil (without spiking), and the results showed that mercury and cadmium were detected by the bioreporters according to the chemical analysis [21].

One bioassay was developed, commercialised, and sold as the BIOMET® kit. This system utilises a set of *C. metallidurans* reporter bacteria, allowing the detection of various pollutants, including zinc, cadmium, copper, chromium, lead [14], and nickel [60]. This test can be used for a quick evaluation of heavy metal bioavailability in polluted soils.

Table 4 Detected heavy metals and detection limits of the different strains of *Escherichia coli*: *Zntlux*, *Arslux*, *Merlux*, and *Coplux*

Heavy metal Tested	Detection limit (standard deviation) (μM)				ES (μM)
	Zntlux	Arslux	Merlux	Coplux	
Cadmium	0.0045 (0.0003)	5.9 (2.3)	0.011 (0.002)	Nd	0.045
Mercury	0.01 (0.005)	Nd	1.2×10^{-7} (1×10^{-7})	Nd	0.005
Arsenic III	28.52 (7.1)	0.256 (0.0014)	15.6 (4.3)	Nd	0.135
Copper	16.92 (2.9)	Nd	Nd	90.5 (11.7)	31.5
Lead	2.2 (0.6)	4.16 (0.8)	Nd	Nd	0.05
Tin	12.95 (4.24)	Nd	Nd	Nd	Nc
Arsenic V	9.32 (1.24)	0.3 (0.06)	12.65 (5.4)	Nd	0.135
Zinc	1.7 (0.62)	Nd	2.3 (0.14)	Nd	Nc
Nickel	4.4 (1.6)	Nd	Nd	Nd	0.34
Cobalt	0.22 (0.014)	Nd	Nd	Nd	Nc
Chromium VI	597.2 (121.3)	Nd	Nd	Nd	1
Chromium III	Nd	Nd	Nd	Nd	1
Silver	Nd	Nd	Nd	2.75 (0.11)	Nc
Iron	4.34 (0.48)	Nd	16.1 (7.6)	Nd	3.5
Manganese	Nd	Nd	Nd	Nd	1

Note ES: European Standards in water according to the European directive 98/83/CE. Nd: metal tested but not detected by the strains. Nc: Not concerned by a standard [32]

5.3 Organometallic Detection

Generally, organometallic compounds are tested in regard to their toxicity on natural or constitutive bioluminescent stains [9]. Detection of organometallic compounds by specific bioassays is not common. Organomercuric compounds, such as methylmercury, have been detected by Merlux, and the detection limit was found to be 2.5 nM for $HgCH_3$ [46].

Among organometallic compounds, another bioassay for the detection of organotin has been developed. Organotin and especially tributyltin (TBT) have been used as antifouling agents in marine paint, wood preservation, and industrial cooling systems. TBT has been considered a major pollutant ever since it was introduced into marine ecosystems. Organotin compounds are classified as 'dangerous priority substances' in the field of water policy (2455/2001/EC, 2001). Detection of organotin requires extraction procedures from the sample and complex chemical methods (chromatographic flame photometric detection or liquid chromatography inductively-coupled plasma mass spectrometry LC-ICP-MS). A bioluminescent *E. coli* specific to TBT was obtained after random insertion of *luxAB* genes from *V. harveyi* [8]. This bacterium detected only TBT and dibutyltin

(DBT), and performance of the bioassay was assessed. The detection limits were found to be 26 μg L^{-1} (0.08 μM) for TBT and 0.03 μg L^{-1} (0.0001 μM) for DBT [17]. The insertion of the *luxAB* gene was studied by Gueuné et al. [23].

Even if this bacterium is very specific to TBT and DBT, its sensitivity is not sufficient for application to marine water samples. A bioassay for TBT detection in marine paint without any extraction procedure was performed. Figure 6 depicts a simple way to control the TBT regulation for the antifouling paint. This bioassay has been applied to real environmental samples from a shipyard. The induction of the bacteria was correlated with the chemical analyses of TBT and DBT [24].

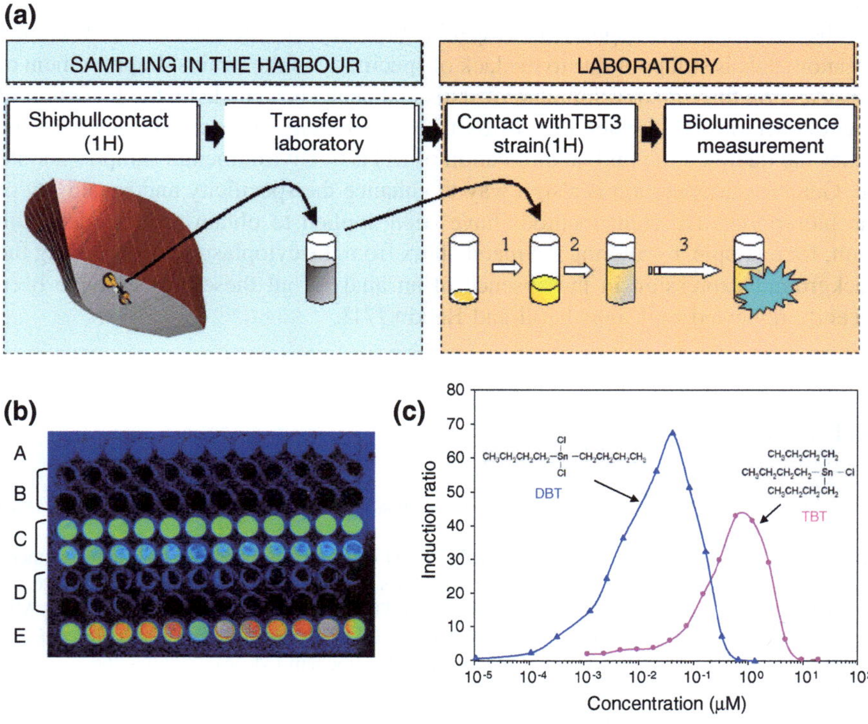

Fig. 6 Example of an application with recombinant bioluminescent bacteria. The bacteria *E. coli* TBT3 (*luxAB* or *luxCDABE* genes) were used in two ways to detect the TBT contained in paint. **a** This diagram illustrates TBT detection in the paint directly using a simple device, a container with suction pads attached. Water with the diffused TBT is sampled and used to induce the *E. coli* TBT3; **b** the second possibility is to introduce the paint into a microtitre plate, after which the recombinant strain, added directly, is induced or not depending on whether the paint contains TBT or DBT (ORACA II ER Hamamastsu CCD camera); **c** the relationship between bioluminescence and the TBT concentration (induction ratio is the ratio between the bioluminescence with TBT versus the background bioluminescence without TBT). *A* control with sea water; *B* paint without TBT; *C* paint with TBT (27 mg kg^{-1}) and DBT (28 mg kg^{-1}); *D* empty wells; *E* control with TBT in seawater (1 μM final concentration) [17, 23, 24]

Another bioassay using a bioluminescent yeast has been developed by Kabiersch et al. based on the interaction with a chimeric human retinoid X receptor for the detection of TBT with a limit of detection of 60 nM [33].

6 Perspectives and Conclusions

Bioassays are widely used in ecotoxicological studies to evaluate the toxicity of environmental samples. Because analytical procedures are time consuming and costly, detection of metals by engineered bacteria could be useful. Moreover, these bioassays provide information on the bioaccessibility and biodisponibility of metals.

The environmental application of specific bioassays for the detection of MTE and organometals is still poor due to the lack of specificity of the strains. Improvement of the detection through development of statistical analysis is made by the induction of a set of bacteria. The greater the sensitivity and specificity of specific bacteria are, the better the qualification and quantification of MTE in environmental samples should be. Genetic manipulation is also a way to enhance the specificity and sensitivity of the bioreporter. Different methods have been applied to obtain a better detection limit, for example, by altering the metal efflux from the cytoplasm or by reducing the background expression in the absence of an analyte; all these aspects have been recently reviewed by Yagur-Kroll and Belkin [71].

References

1. Atkins PW, Jones LL (1997) Chemistry: molecules, matter, and change, third. W.H. Freeman and Co., New York
2. Audry S, Schäfer J, Blanc G, Jouanneau JM (2004) Fifty-year sedimentary record of heavy metal pollution (Cd, Zn, Cu, Pb) in the lot river reservoirs (France). Environ Pollut 132:413–426
3. Baldwin TO, Berends T, Bunch TA, Holzman TF, Rausch SK, Shamansky L, Treat ML, Ziegler MM (1984) Cloning of the luciferase structural genes from *Vibrio harveyi* and expression of bioluminescence in *Escherichia coli*. Biochem (Mosc) 23:3663–3667
4. Baumann B, van der Meer JR (2007) Analysis of bioavailable arsenic in rice with whole cell living bioreporter bacteria. J Agric Food Chem 55:2115–2120
5. Bjerketorp J, Håkansson S, Belkin S, Jansson JK (2006) Advances in preservation methods: keeping biosensor microorganisms alive and active. Curr Opin Biotechnol 17:43–49
6. Boyanapalli R, Bullerjahn GS, Pohl C, Croot PL, Boyd PW, McKay RML (2007) Luminescent whole-cell cyanobacterial bioreporter for measuring Fe availability in diverse marine environments. Appl Environ Microbiol 73:1019–1024
7. Branco R, Cristóvão A, Morais PV (2013) Highly sensitive, highly specific whole-cell bioreporters for the detection of chromate in environmental samples. PLoS ONE 8:e54005
8. Briscoe SF, Diorio C, DuBow MS (1996) Luminescent biosensors for the detection of tributyltin and dimethyl sulfoxide and the elucidation of their mechanisms of toxicity. Environ Biotechnol, pp 645–655

9. Bundy JG, Wardell JL, Campbell CD, Killham K, Paton GI (1997) Application of bioluminescence-based microbial biosensors to the ecotoxicity assessment of organotins. Let Appl Microbiol 25:353–358
10. Cai J, DuBow MS (1997) Use of a luminescent bacterial biosensor for biomonitoring and characterization of arsenic toxicity of chromated copper arsenate (CCA). Biodegradation 8:105–111
11. Charrier T, Durand MJ, Jouanneau S, Dion M, Pernetti M, Poncelet D, Thouand G (2011) A multi-channel bioluminescent bacterial biosensor for the on-line detection of metals and toxicity, part I: design and optimization of bioluminescent bacterial strains. Anal Bioanal Chem 400:1051–1060
12. Checa SK, Zurbriggen MD, Soncini FC (2012) Bacterial signaling systems as platforms for rational design of new generations of biosensors. Curr Opin Biotechnol 23:766–772
13. Christophoridis C, Dedepsidis D, Fytianos K (2009) Occurrence and distribution of selected heavy metals in the surface sediments of Thermaikos Gulf, N. Greece: assessment using pollution indicators. J Hazard Mater 168:1082–1091
14. Corbisier P, Ji G, Nuyts G, Mergeay M, Silver S (1993) luxAB gene fusions with the arsenic and cadmium resistance operons of staphylococcus aureus plasmid pI258. FEMS Microbiol Lett 110:231–238
15. Corbisier P, van der Lelie D, Borremans B, Provoost A, de Lorenzo V, Brown NL, Lloyd JR, Hobman JL, Csöregi E, Johansson G, Mattiasson B (1999) Whole cell- and protein-based biosensors for the detection of bioavailable heavy metals in environmental samples. Anal Chim Acta 387:235–244
16. Dunlap PV (2009) Bioluminescence, microbial. In: Schaechter M (ed) Encyclopedia of microbiology. Elsevier, Oxford, pp 45–61
17. Durand MJ, Thouand G, Dancheva-Ivanova T, Vachon P, DuBow M (2003) Specific detection of organotin compounds with a recombinant luminescent bacteria. Chemosphere 52:103–111
18. Eltzov E, Marks RS (2011) Whole-cell aquatic biosensors. Anal Bioanal Chem 400:895–913
19. Erbe JL, Adams AC, Taylor KB, Hall LM (1996) Cyanobacteria carrying an smt-lux transcriptional fusion as biosensors for the detection of heavy metal cations. J Ind Microbiol 17:80–83
20. Flynn HC, Mc Mahon V, Diaz GC, Demergasso CS, Corbisier P, Meharg AA, Paton GI (2002) Assessment of bioavailable arsenic and copper in soils and sediments from the antofagasta region of northern Chile. Sci Total Environ 286:51–59
21. Foucault Y, Durand MJ, Tack K, Schreck E, Geret F, Leveque T, Pradere P, Goix S, Dumat C (2013) Use of ecotoxicity test and ecoscores to improve the management of polluted soils: case of a secondary lead smelter plant. J Hazard Mater 246–247:291–299
22. Girotti S, Ferri EN, Fumo MG, Maiolini E (2008) Monitoring of environmental pollutants by bioluminescent bacteria. Anal Chim Acta 608:2–29
23. Gueuné H, Durand MJ, Thouand G, DuBow MS (2008) The ygaVP genes of *Escherichia coli* form a tributyltin-inducible operon. Appl Environ Microbiol 74:1954–1958
24. Gueuné H, Thouand G, Durand MJ (2009) A new bioassay for the inspection and identification of TBT-containing antifouling paint. Mar Pollut Bull 58:1734–1738
25. Guzzo A, Diorio C, DuBow MS (1991) Transcription of the *Escherichia coli* fliC gene is regulated by metal ions. Appl Environ Microbiol 57:2255–2259
26. Guzzo A, DuBow MS (1991) Construction of stable, single-copy luciferase gene fusions in *Escherichia coli*. Arch Microbiol 156:444–448
27. Hakkila K, Maksimow M, Karp M, Virta M (2002) Reporter genes lucFF, luxCDABE, gfp, and dsred have different characteristics in whole-cell bacterial sensors. Anal Biochem 301:235–242
28. Lemire JA, Harrison JJ, Turner RJ (2013) Antimicrobial activity of metals: mechanisms, molecular targets and applications. Nat Rev Microbiol 11:371–384
29. Hynninen A, Tönismann K, Virta M (2010) Improving the sensitivity of bacterial bioreporters for heavy metals. Bioeng Bugs 1:132–138

30. Ivask A, Green T, Polyak B, Mor A, Kahru A, Virta M, Marks R (2007) Fibre-optic bacterial biosensors and their application for the analysis of bioavailable Hg and As in soils and sediments from Aznalcollar mining area in Spain. Biosens Bioelectron 22:1396–1402
31. Ivask A, Rolova T, Kahru A (2009) A suite of recombinant luminescent bacterial strains for the quantification of bioavailable heavy metals and toxicity testing. BMC Biotechnol 9:41
32. Jouanneau S, Durand MJ, Courcoux P, Blusseau T, Thouand G (2011) Improvement of the identification of four heavy metals in environmental samples by using predictive decision tree models coupled with a set of five bioluminescent bacteria. Environ Sci Technol 45:2925–2931
33. Kabiersch G, Rajasarkka J, Tuomela M, Hatakka A, Virta M, Steffen K (2013) Bioluminescent yeast assay for detection of organotin compounds. Anal Chem 85:5740–5745
34. Kaushik A, Kansal A, Santosh Meena, Kumari S, Kaushik CP (2009) Heavy metal contamination of river Yamuna, Haryana, India: assessment by metal enrichment factor of the sediments. J Hazard Mater 164:265–270
35. Khang Y, Yang Z, Burlage R (1997) Measurement of iron-dependence of pupA promoter activity by a pup-lux bioreporter. J Microbiol Biotechnol 7:352–355
36. Klein JS, Lewinson O (2011) Bacterial ATP-driven transporters of transition metals: physiological roles, mechanisms of action, and roles in bacterial virulence. Metallomics 3:1098–1108)
37. Kim EH, Nies DH, McEvoy MM, Rensing C (2011) Switch or funnel: how RND-type transport system control periplasmic metal homeostasis. J Bacteriol 193:2381–2387
38. Ma Z, Jacobsen FE, Giedroc DP (2009) Coordination chemistry of bacterial metal transport and sensing. Chem Rev 109:4644–4681
39. Magrisso S, Erel Y, Belkin S (2008) Microbial reporters of metal bioavailability. Microb Biotechnol 1:320–330
40. Maier RM, Pepper IL, Gerba CP (2000) Environmental microbiology. London Academic Press, 585 p
41. McElroy WD, Ballentine R (1944) The mechanism of bioluminescence. Proc Natl Acad Sci U S A 30:377–382
42. Meighen EA (1993) Bacterial bioluminescence: organisation, regulation, and application of the lux genes. FASEB J 7:1016–1022
43. Meighen EA (1994) Genetics of bacterial bioluminescence. Annu Rev Genet 28:117–139
44. Mergeay M, Monchy S, Vallaeys T, Auquier V, Benotmane A, Bertin P, Taghavi S, Dunn J, van der Lelie D, Wattiez R (2003) Ralstonia metallidurans, a bacterium specifically adapted to toxic metals: towards a catalogue of metal-responsive genes. FEMS Microbiol Rev 27:385–410
45. Merulla D, Buffi N, Beggah S, Truffer F, Geiser M, Renaud P, van der Meer JR(2013) Bioreporters and biosensors for arsenic detection: biotechnological solutions for a world-wide pollution problem. Curr Opin Biotechnol 24(3):534–41
46. Ndu U, Mason RP, Zhang H, Lin S, Visscher PT (2012) Effect of inorganic and organic ligands on the bioavailability of methylmercury as determined by using a mer-lux bioreporter. Appl Environ Microbiol 78:7276–7282
47. Nies DH (1999) Microbial heavy-metal resistance. Appl Microbiol Biotechnol 51:730–750
48. Peca L, Kós PB, Máté Z, Farsang A, Vass I (2008) Construction of bioluminescent cyanobacterial reporter strains for detection of nickel, cobalt and zinc. FEMS Microbiol Lett 289:258–264
49. Pepi M, Reniero D, Baldi F, Barbieri P (2006) A comparison of MER:LUX whole cell biosensors and moss, a bioindicator, for estimating mercury pollution. Water Air Soil Pollut 173:163–175
50. Qiao Y, Yang Y, Zhao J, Tao R, Xu R (2013) Influence of urbanization and industrialization on metal enrichment of sediment cores from Shantou Bay, South China. Environ Pollut 182:28–36
51. Ren S, Frymier PD (2004) Reducing bioassay variability by identifying sources of variation and controlling key parameters in assay protocol. Chemosphere 57:81–90

52. Riether KB, Dollard MA, Billard P (2001) Assessment of heavy metal bioavailability using *Escherichia coli* zntAp:lux and copAp:lux-based biosensors. Appl Microbiol Biotechnol 57:712–716
53. Selifonova O, Burlage R, Barkay T (1993) Bioluminescent sensors for detection of bioavailable Hg(II) in the environment. Appl Environ Microbiol 59:3083–3090
54. Shin HJ (2011) Genetically engineered microbial biosensors for in situ monitoring of environmental pollution. Appl Microbiol Biotechnol 89:867–877
55. Sørensen SJ, Burmølle M, Hansen LH (2006) Making bio-sense of toxicity: new developments in whole-cell biosensors. Curr Opin Biotechnol 17:11–16
56. Stocker J, Balluch D, Gsell M, Harms H, Feliciano J, Daunert S, Malik KA, van der Meer JR (2003) Development of a set of simple bacterial biosensors for quantitative and rapid measurements of arsenite and arsenate in potable water. Environ Sci Technol 37:4743–4750
57. Summers AO (2009) Damage control: regulating defenses against toxic metals and metalloids. Curr Opin Microbiol 12:138–144
58. Tauriainen S, Karp M, Chang W, Virta M (1998) Luminescent bacterial sensor for cadmium and lead. Biosens Bioelectron 13:931–938
59. Thouand G, Durand MJ (2013) Chapter 13, bacteria in ecotoxicology: recombinant luminescent bacteria. In: Blaise C, Ferard JF (eds) Encyclopedia of aquatic ecotoxicology. Springer, Berlin, pp 137–150
60. Tibazarwa C, Corbisier P, Mench M, Bossus A, Solda P, Mergeay M, Wyns L, van der Lelie D (2001) A microbial biosensor to predict bioavailable nickel in soil and its transfer to plants. Environ Pollut 113:19–26
61. Tibazarwa C, Wuertz S, Mergeay M, Wyns L, van Der Lelie D (2000) Regulation of the cnr cobalt and nickel resistance determinant of *Ralstonia eutropha* (*Alcaligenes eutrophus*) CH34. J Bacteriol 182:1399–1409
62. Trang PTK, Berg M, Viet PH, Mui NV, van der Meer JR (2005) Bacterial bioassay for rapid and accurate analysis of arsenic in highly variable groundwater samples. Environ Sci Technol 39:7625–7630
63. Trögl J, Chauhan A, Ripp S, Layton AC, Kuncová G, Sayler GS (2012) Pseudomonas fluorescens HK44: lessons learned from a model whole-cell bioreporter with a broad application history. Sensors 12:1544–1571
64. Ulitzur S, Goldberg I (1977) Sensitive, rapid, and specific bioassay for the determination of antilipogenic compounds. Antimicrob Agents Chemother 12:308–313
65. Van der Meer JR, Belkin S (2010) Where microbiology meets microengineering: design and applications of reporter bacteria. Nat Rev Microbiol 8:511–522
66. Verma N, Singh M (2005) Biosensors for heavy metals. Biometals Int J Role Met Ions Biol Biochem Med 18:121–129
67. Wenfeng S, Gooneratne R, Glithero N, Weld RJ, Pasco N (2013) Appraising freeze-drying for storage of bacteria and their ready access in a rapid toxicity assessment assay. Appl Microbiol Biotechnol 97:10189–10198
68. World Health Organization (2008) Guidelines for drinking-water quality—volume 1: recommendations, third
69. Xu T, Close DM, Sayler GS, Ripp S (2013) Genetically modified whole-cell bioreporters for environmental assessment. Ecol Indic 28:125–141
70. Yagi K (2007) Applications of whole-cell bacterial sensors in biotechnology and environmental science. Appl Microbiol Biotechnol 73:1251–1258
71. Yagur-Kroll S1, Belkin S (2014) Molecular manipulations for enhancing luminescent bioreporters performance in the detection of toxic chemicals. Adv Biochem Eng Biotechnol 145:137–149

Adv Biochem Eng Biotechnol (2016) 154: 101–116
DOI 10.1007/10_2015_333

Published Online: 17 October 2015

Main Technological Advancements in Bacterial Bioluminescent Biosensors Over the Last Two Decades

S. Jouanneau, M.J. Durand, A. Lahmar and G. Thouand

Abstract Environmental quality assessment is an extensive field of research due to the permanent increase of the stringency imposed by the legislative framework. To complete the wide panel of measurement methods, essentially based on physico-chemical tools, some scientists focused on the development of alternative biological methods such as those based on the use of bioluminescent bacteria biosensors. The first report dedicated to the development of such biosensors dates back to 1967 and describes an analytical system designed to address the problem of air toxicity assessment. Nevertheless the available technologies in the photosensitive sensors field were not mature enough and, as a result, limited biosensor development possibilities. For about 20 years, the wide democratisation of photosensors coupled with advances in the genetic engineering field have allowed the expansion of the scope of possibilities of bioluminescent bacterial biosensors, allowing a significant emergence of these biotechnologies. This chapter retraces the history of the main technological evolutions that bacterial bioluminescent biosensors have known over the last two decades.

S. Jouanneau (✉) · M.J. Durand · A. Lahmar · G. Thouand
University of Nantes, UMR CNRS GEPEA 6144, 18 Bd Gaston Defferre, Nantes 85000, La Roche sur Yon, France
e-mail: Sulivan.jouanneau@univ-nantes.fr

Graphical Abstract

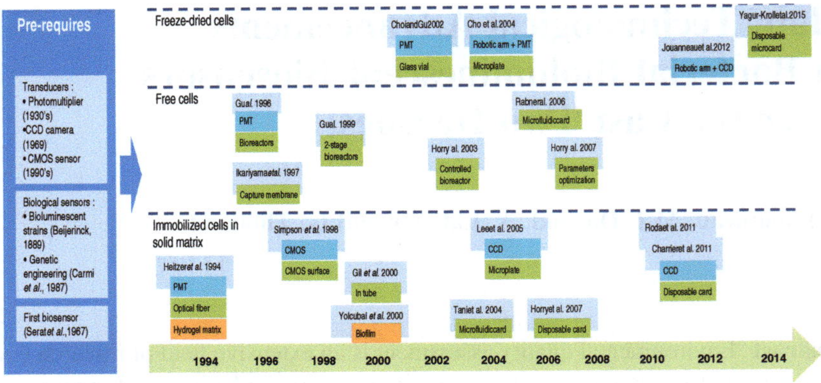

Keywords Bioluminescent bacterial biosensor · Specific detection · Technological evolution · Toxicity assessment

Contents

1 Introduction

A biosensor is defined according to the International Union of Pure and Applied Chemistry as "A device that uses specific biochemical reactions mediated by isolated enzymes, immunosystems, tissues, organelles or whole cells to detect chemical compounds usually by electrical, thermal or optical signals." They differ from bioassays because biosensors incorporate direct spatial contact between the biological recognition element and the transducer [1].

These systems were designed as a novel approach for environmental metrology that would be complementary to the physicochemical methods used to evaluate the impact of pollutants on the biology of organisms (i.e., toxicity and bioavailable fractions). The biosensors developed over the last few decades are primarily used in three business sectors: environmental monitoring, food processing, and the medical

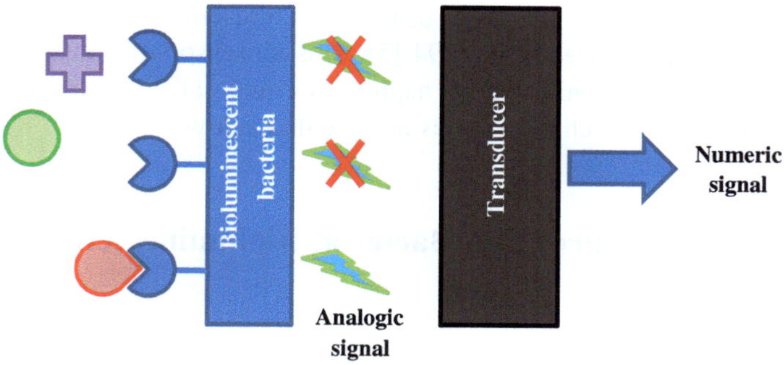

Fig. 1 Principle of a bacterial bioluminescent biosensor

industry. The benefits of these biological tools include their user-friendliness, moderate cost, and short response time.

In this chapter, we focus only on bacterial bioluminescent biosensors (BBBs). These biosensors are based on living bacterial cells and are particularly interesting for the assessment of the bioavailable fraction of specific compounds and their toxicological impacts [2]. A BBB consists of a bioluminescent bacterial bioelement for recognition and a photosensitive transducer (Fig. 1). In the presence of the detected compounds, the bacteria emit a bioluminescent signal (increase or decrease) relative to the exposition intensity (i.e., toxicity level and concentration). This biological signal is recorded by the photosensitive transducer and converted into a numeric signal. Two types of bioelements are used according to the desired procedure:

- *Toxicological assessment.* For this purpose, the bioelements emit a significant signal of bioluminescence. This procedure involves the use of wild bioluminescent bacterial strains or strains that have been modified with a genetic construct containing the bioluminescence operon (*lux* or *luc*) under the control of a constitutive promoter. The decrease in bioluminescence emitted by the bacterial cells is recorded to assess the toxicity level in the tested sample.
- *Specific detection of chemical compounds*. In this case, the bioelements contain the bioluminescence operon under the control of an inducible promoter. The bioluminescence level emitted by the bacterial cells increases significantly in the presence of specific compounds, which allows the user to quantitate the contamination of the analysed environment.

The first report of the utilisation of bioluminescent bacteria for toxicity monitoring was the publication in 1959 by Strehler [3]. Based on this work, the first biosensor based on immobilised wild bioluminescent bacteria (*Photobacterium phosphoreum*) was developed and described in 1967 by Serat et al. [4]. This device was designed to address the problem of air toxicity assessment. Since this first device, biosensors have been subjected to many technological improvements. This chapter should provide a better understanding of the main evolutions implemented in the biosensor domain. This chapter is entirely devoted to the technical aspects and the measurement strategies of the biosensors described in the literature over the last two decades.

During this period, the first BBB using these bioelements (inducible bacteria) was described by Heitzer et al. in 1994 [5]. This biosensor was developed for the environmental online monitoring of naphthalene and salicylate in aquatic environments. The technical characteristics are detailed below.

2 Main Architectures of a Bacterial Bioluminescent Biosensor

Technical advances in the field of physics, particularly those concerning photosensitive sensors, have greatly contributed to the development of the biosensor [6]. Indeed, the increased scope of these sensors has allowed a reduction in the cost of these technologies and a simplification of their implementation, which has provided access to new development perspectives.

In parallel, the scientific and technological progress in genetic engineering has allowed the development of new detection tools such as biological probes and bacterial bioelements [7]. In this case, a DNA insert (bioluminescent operon (*lux* or *luc*) under the control of a constitutive or inducible promoter) is added to the genetic heritage of the host strain (insertion of a plasmid or chromosomal modification). This genetic modification allows the cells to change the bioluminescence emission level in reaction to environmental stimuli (i.e., stress, toxicity, or the presence of specific compounds).

By combining these two technological advances, Heitzer et al. [5] designed an inducible biosensor in 1994 based on immobilised cells in direct contact with the optical transducer: a photomultiplier (PMT) via an optical fibre to detect specific chemicals (i.e., naphthalene and salicylate) in aquatic environments. This precursor opened a pathway for many other innovative technologies in the field of environmental monitoring using bacterial bioluminescent biosensors, as shown in Fig. 2 and described below.

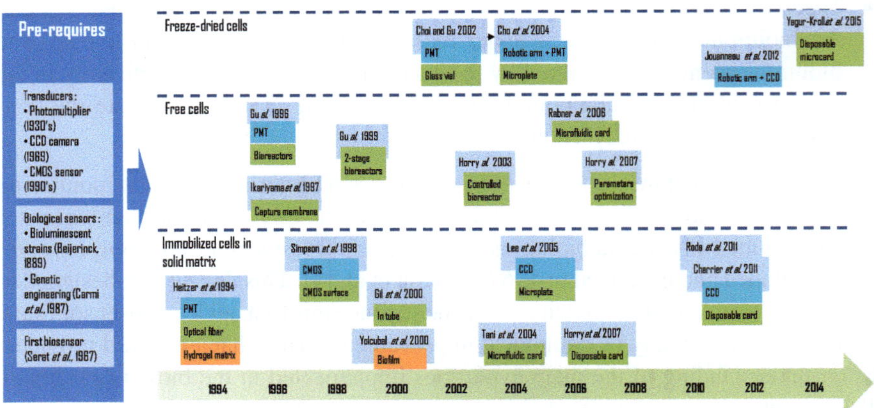

Fig. 2 Evolution of bioluminescent bacterial biosensor over the last two decades [4, 14, 15]

2.1 *Biosensors Based on Immobilised Bioluminescent Bacteria*

The immobilisation of bioluminescent cells was chosen to restrain their growth in a defined area, to limit their release into the environment (notably due to issues with genetically modified microorganisms), and to reuse the bioelements for as long as possible. Several types of matrices were studied for this purpose. This aspect has been widely treated in the literature [8–10] and is thus not addressed in this chapter.

The first strategy consisted of the immobilisation of cells in a hydrogel matrix (a type of polymer network) on an optical fibre to ensure direct contact between the bacteria and the photosensitive sensor. Several configurations were developed according to the desired applications (i.e., water or gas monitoring) and the type of optical sensors [i.e., PMT, charge-coupled device (CCD), or complementary metal oxide semiconductor (CMOS)].

2.1.1 Type of Transducer: Photomultipliers (PMT)

The PMT technologies appeared in the 1930s [11] following the work of Iams and Salzberg [12] and Zworykin et al. [13]. These photodetectors were extremely light-sensitive sensors that multiplied the current produced by incident light using a succession of dynodes (electrodes responsive to signal amplification), which enabled individual photons to be detected when the light intensity was very low.

Most biosensors using this technology function as bioluminescent sensors and are based on the same schematic. Bioluminescent bacterial cells are placed in contact with an optical fibre tip that is responsible for transferring the bioluminescence emitted by the cells to the PMT. Therefore, it is possible to quantify variations in bioluminescence intensity based on the exposition of the analysed sample.

Biological Probes Based on Cells Immobilised on the Optical Fibre Tip

The first biosensor developed by Heitzer et al. [5] was related to physicochemical immersion probes such as dissolved oxygen sensors or pH probes. It was based on the immobilisation of cells in a hydrogel membrane that was composed of alginate. After the polymerisation of the hydrogel membrane with the bacteria, the probe was placed at the extremity of an optical fibre (Fig. 3). Using this approach, control of the immobilised cell number and the membrane thickness was facilitated; however, this approach involves placing the bacterial membrane onto the optical fibre tip to ensure contact between them. To overcome this additional step and allow for miniaturisation, Polyak et al. [16] proposed an alternative method with a simplified implementation based on the multilayer immobilisation of alginate on the optical fibre. In this case, the optical fibre tip is immersed in several successive baths of polymer with bacteria to form a membrane consisting of the superimposition of fine

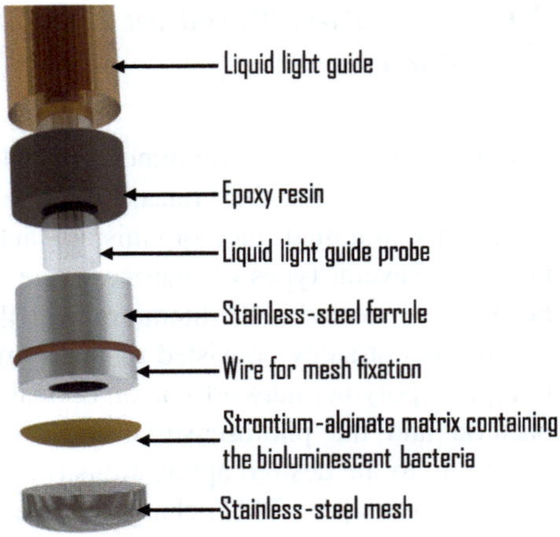

Fig. 3 Biosensor designed by Heitzer et al. [5] and based on entrapped cells on an optical fibre tip

layers of polymer containing bacterial cells [16, 17]. This biosensor architecture was also used by Hakkila et al. [18] to design a multitarget biosensor based on several specific bioelements.

Bacterial Bioluminescent Biosensors Without Direct Contact Between the Bacteria and Transducer

The first representative of this biosensor family was described in 1967 by Serat et al. [4] and was proposed to address the problem of air toxicity assessment. Based on this strategy, Cheol Gil et al. [19] proposed a biosensor for the specific detection of benzene in the atmospheric compartment. In contrast to the biosensors described in the previous paragraph, the bacterial cells were not entrapped directly on the optical fibre tip but were encased in a transparent support composed of polypropylene [19] (Fig. 4a). This configuration allowed for the quick and easy replacement of the disposable bioelements. Nevertheless, the authors stressed that this approach was severely limited by the diffusion of gas into the hydrogel matrix. Consequently, Gu et al. [20] and Chang et al. [21] proposed the addition of small glass beads into a hydrogel matrix containing bacterial cells to enhance the porosity of the matrix and facilitate gas diffusion. Using a similar strategy, Yolcubal et al. [22] developed a biosensor with the addition of sand in place of the glass beads. For a similar application, Valdman and Gutz [23] proposed another alternative method based on slides (1.5 mm thick) of hydrogel that contained entrapped bacterial cells. According to the authors, this configuration ensured a large exchange surface between the bacteria and the analysed gas.

To address the problem of online water monitoring, Horry et al. [24] proposed a biosensor based on a disposable card containing the entrapped bacteria. This configuration allowed for the exposure of the cells to several successive samples and for a rapid and simple replacement of bioelements after extended utilisation. Moreover, because the temperature is an essential parameter of the biological

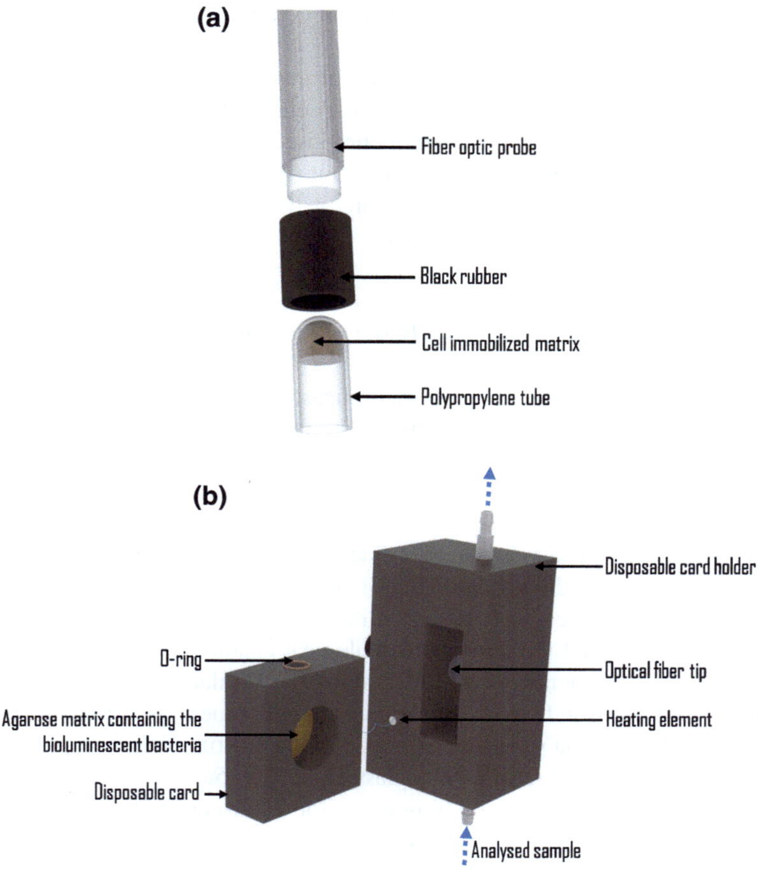

Fig. 4 Biosensors based on disposable module containing bioluminescent bacterial cells. **a** Cheol Gil et al. [19]; **b** Horry et al. [24]

detection process, this system integrates a temperature regulator consisting of a heating cartridge, temperature probe, and controller (Fig. 4b).

2.1.2 Type of Transducer: CCD (Charged-Coupled Device) or CMOS (Complementary Metal–Oxide Semiconductor)

The CCD (developed in 1969 by Willard S. Boyle and George E. Smith) and CMOS (developed by Mendis et al. in the 1990s [25]) technologies belong to the category of image sensors. Indeed, these technologies, which are newer than the PMT technology, were developed to perform photographic captures and are widely used in digital cameras. This second type of light detector was used to monitor variations in the bioluminescence of immobilised bacteria. Nevertheless, these technologies require specific configurations in the biosensors.

Bacterial Bioluminescent Biosensors Based on CMOS Sensors

Though this optical technology is more recent than the CCD sensors, it has been integrated into bacterial biosensors by Simpson et al. [26] since 1998 because they serve as inexpensive sensors. The authors proposed a new concept for a biosensor: the bioluminescent bioreporter integrated circuit (BBIC), in which the immobilised bacteria are inserted into an integrated circuit that is in direct contact with the photodetector as an electronic compound.

Bacterial Bioluminescent Biosensors Based on the CCD Camera

The strategy of these biosensors consists of measuring the bioluminescence emitted by immobilised bacteria with a CCD camera placed above or below the immobilised cells to obtain an image. The first biosensor using this optic technology was described by Lee et al. in 2005 [27]. In this study, the inducible bioluminescent bacteria (20 recombinant strains) were immobilised in an agar hydrogel in a 384-well plate. The plate was placed inside a light-tight box integrating the CCD camera. In addition to this technological innovation, the authors proposed a new strategy for toxicity assessment based on the use of several recombinant bacterial strains, which increased the level of toxicological characterisation in the analysed samples. In 2007, Sakaguchi et al. [28] proposed a similar biosensor based on the use of the CCD sensor from a commercial digital camera as a bioluminescence sensor. Here, the bacteria were immobilised inside a matrix of alginate polymer that was deposited on the microwells of a glass slide (Fig. 5a).

To simplify the utilisation of biosensors for environmental conditions, Roda et al. [29] designed a portable device in which cells were immobilised in an interchangeable cartridge. Using a fibre-optic taper, the bacteria were placed in direct contact with the cooled CCD sensor, which allowed for the precise measurement of the bioluminescence signal while still allowing for the regulation of the temperature of the bacteria.

The architecture of the biosensors described above requires replacing the support containing the bacteria after each analysis (the disposable part) or a manual washing step. Consequently, it is necessary to repeat all bacterial preparation operations (i.e., growth and immobilisation) before performing another experiment. To reduce this technical constraint, Charrier et al. [30] proposed an online biosensor based on a disposable card containing immobilised cells (Fig. 5b) that is continuously fed with liquids (samples or regenerating media). During an analysis cycle, the cells are exposed to the sample premixed with a nutrient solution. After each analysis cycle, the bacteria are automatically regenerated with the nutrient solution prior to the next analysis.

The biosensor designed by Tani et al. in 2004 [31] was based on the same strategy as the biosensor designed by Charrier et al. [30]: the bacteria were immobilised in a hydrogel matrix of agarose and fed continuously with nutrients to allow cell regeneration after exposure to pollutants. The main difference with

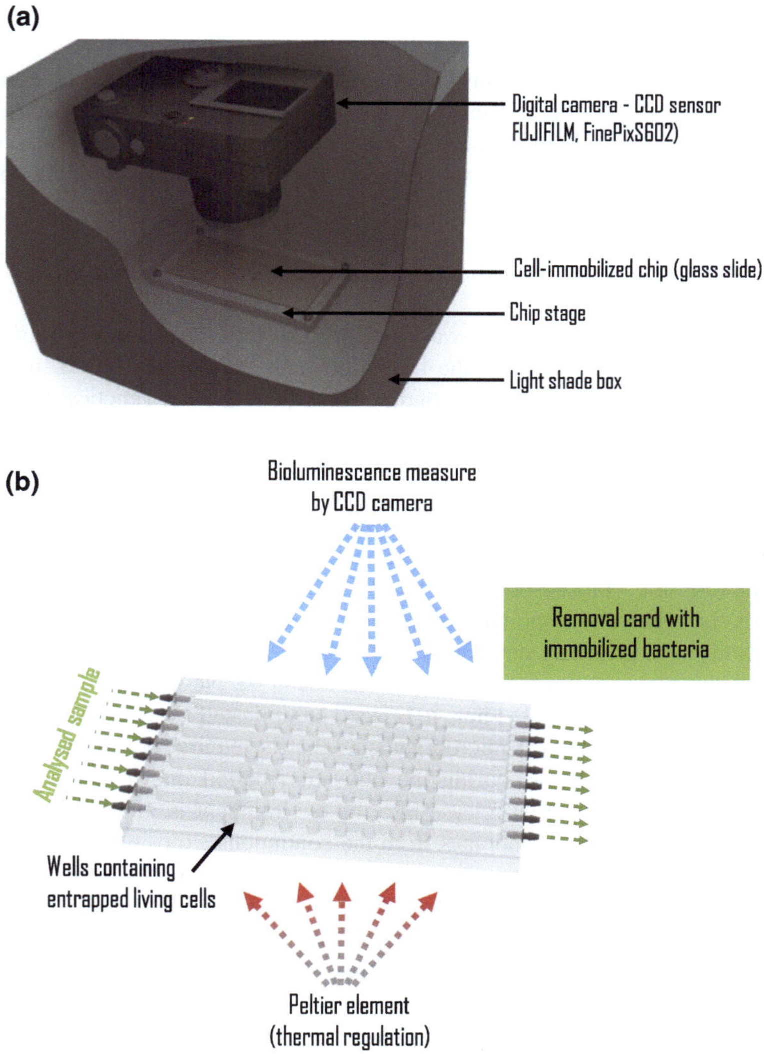

Fig. 5 Immobilised bacterial biosensors with CCD sensor. **a** Sakaguchi et al. [28]; **b** Charrier et al. [30]

Charrier's biosensor concerns the scale. The disposable card was replaced by a much smaller support (a microfluidic card) manufactured in polydimethylsiloxane (PDMS). This technological advance should allow for an increase in the bioelement number in the same space without oversizing the biosensor.

In practice, it appears that these approaches are complex to implement for environmental samples [32]. The multiplication of immobilised bacterial cells in the polymer matrix tends to result in the formation of a biofilm that colonizes the surface of the support and disrupts bioluminescence production due to nutrient availability and the physiological state. This microbial growth also contributes to the release of the bioluminescent cells into the environment. Consequently, all liquids that were in contact with the immobilised bioelements must be collected. Moreover, the continuous feeding with nutrients promotes the growth of environmental microbial flora in the biosensor. The direct consequence of this phenomenon is the consumption of the nutrients necessary for bioluminescence production.

2.1.3 Advantages and Drawback of These Immobilised Biosensors

The use of immobilised bacteria was developed to fulfil three main requirements: first, to assure contact between the bioluminescent bacterial cells and the photosensor (PMT, CCD, or CMOS); second, to limit the release of genetically modified bacteria into the environment; and third, to reuse the bioelements after an automatic washing step and therefore to increase their life span. However, this approach may be problematic for in situ applications. Indeed, immobilised bacteria are living cells that may change over time (i.e., changes in growth and metabolic status), thereby potentially modifying the biological response.

For "one-shot" biosensors (Fig. 5a, b), the cost of preparation for only one analysis is relatively important in terms of time and technicality. The other biosensors were primarily developed to perform several successive measurements to make the preparation cost-effective (Figs. 3 and 4). Nevertheless, this configuration poses problems [32]. First, it is very complicated to maintain the perfect stability of the immobilised microorganisms that is required to ensure the reproducibility of measurements during the utilisation period due to cellular growth. Second, the risk of bacterial release into the environment is important due to the erosion of the hydrogel membranes from the continuous exposition of the samples (i.e., the transfer of antibiotic resistance genes to the environmental microflora). Third, the response of the bacterial membranes is directly influenced by the previous expositions (i.e., toxicity and disturbances due to environmental microflora). Finally, limitations in nutrient diffusion (e.g., O_2 and, sources of C/N/P) into the different layers of the polymer membrane are likely to disrupt the bioluminescence signal; this finding was highlighted by Affi et al. [33].

2.2 *Bacterial Bioluminescent Biosensors Based on Free Cells*

The main interest in the development of BBBs based on free cells is the rapid diffusion of the different analysed compounds through the liquid matrix containing the bioelements, in contrast to biosensors with immobilised cells.

2.2.1 Bioreactor Configuration

The first strategy described in this chapter was proposed by Gu et al. [34] in 1996. This device was dedicated to monitoring water toxicity using wild bioluminescent bacteria. In term of its conception, it resembles a simplified culture bioreactor lacking a control probe (Fig. 6a). The bioluminescence monitoring was performed with a PMT connected to the bioreactor via an optical fibre. Nevertheless, the lack of control in this biosensor did not improve the stability of the cells.

Following this work, in 1999, Gu et al. [35] developed a multichannel version of this biosensor [36, 37]. This biosensor consisted of two successive minibioreactors: the first was specifically dedicated to maintaining the bacterial cells in a stable state and allowed a continuous supply to the second reactor, which corresponded to the analytical part of the biosensor. The stability of the bacterial cells is not guaranteed

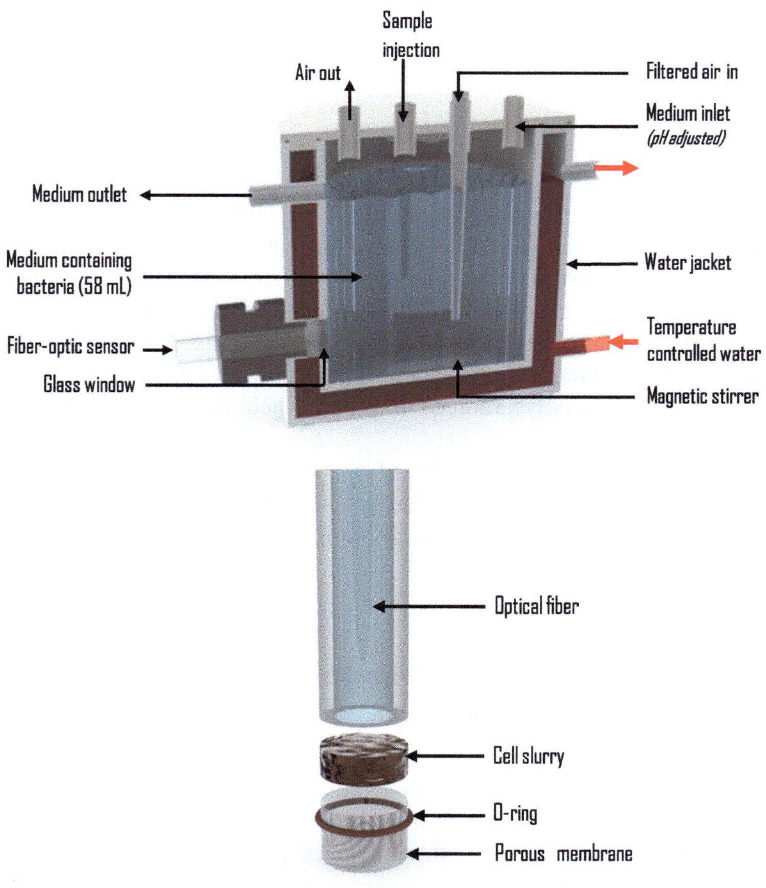

Fig. 6 Examples of biosensor's architecture based on free cells. **a** Gu et al. [34]; **b** Ikariyama et al. [39]

due to the lack of a control probe. However, with this new architecture the bacterial cells are continuously renewed in the analytical reactor. Consequently, the influence of previous expositions is limited.

To compensate for this lack of microbial monitoring (i.e., physiological state, available dissolved oxygen, and temperature), Horry et al. [38] proposed a monoreactor biosensor including a micro-probe set to measure pH, temperature, dissolved oxygen, and cellular density. This addition significantly improved the monitoring of the physiological status of the bacterial cells in the reactor but added greater complexity in its conception. The samples are directly injected into the bioreactor in this biosensor and create a risk of disrupting the bacterial cells and modifying their reactions in the next analysis.

The association of these two technologies (i.e., the multireactor concept of Gu and Gil [36] and the controlled bioreactor proposed by Horry et al. [38]) represents an interesting compromise that allows the reliable monitoring of toxicity by bioluminescent bacteria. Nevertheless, the implementation of these bioreactors remains complex; consequently, they are only barely transferable to in situ monitoring applications.

2.2.2 Probe Configurations

The biosensor proposed by Ikariyama et al. in 1997 [39] was similar to the biosensor based on immobilised cells developed by Heitzer et al. (Cf. previous section), with the exception of the cell hold mode. The cells were maintained in the liquid phase in this biosensor, which allowed for the rapid transfer of the analysed compounds from sample to bacteria for quicker analysis. For this purpose, the bacteria are directly entrapped on the optical fibre tip with a porous membrane (e.g., a dialysis membrane or polycarbonate film) that is held in place on the optical fibre using an o-ring (Fig. 6b).

In the context of the development of a portable micro total analysis system (Lab-On-Chip), Rabner et al. [40] designed a microbiosensor using microfluidic technologies to ensure the mixture of cells with the sample and to contribute to the bioluminescence measurement.

2.3 Bacterial Bioluminescent Biosensors Based on Freeze-Dried Cells

To transfer this technology to industry applications, it was essential to propose a "plug and play" strategy without a requirement for bacterial cultivation, an immobilisation procedure, or any other microbial preparation steps. Consequently, several scientists evaluated the conservation of microorganisms for long-term use. Among the conservation methods of microorganisms usually used, freeze-drying seemed to be a pertinent compromise.

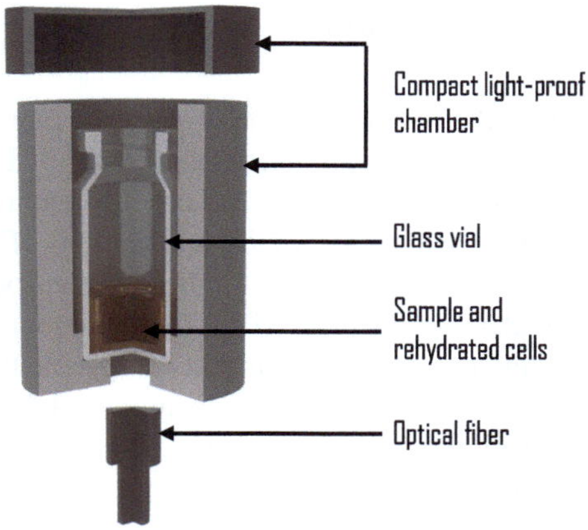

Fig. 7 Example of biosensor based on freeze-dried cells, Choi and Gu [41]

However, the implementation of this storage mode of cells requires a supplementary step of preparation that consists of dehydrating the biological cells in sublimation conditions with water (–50 °C or–85 °C) and under very low pressure (near 0.05 mbar). Protected from light, humidity, and oxidation, the lyophilised cells are stable and can be stored for a long period of time without evolution.

The first listed bacterial biosensor based on freeze-dried cells was proposed by Choi and Gu in 2002 [41]. This portable biosensor consisted of three parts: a glass vial containing the freeze-dried bacteria, a compact lightproof chamber, and a luminometer (optical fibre and PMT; Fig. 7). With this configuration, the user's operations were reduced to rehydrating the cells 30 min prior to performing the analysis, adding the samples, and reading the bacterial bioluminescence; no additional preparation was required.

To allow the high-throughput screening of water toxicity, Cho et al. [42] developed an automated continuous toxicity test system in which the bacteria were freeze-dried in 384-well microplates. All of the manual steps performed in the biosensor of Choi and Gu were replaced by a robotic arm. For the sake of simplification of the bioluminescence measure, Jouanneau et al. [32] exchanged the PMT with a less expensive means: a CCD camera coupled with picture analysis software. This evolution greatly simplified the technical aspects of the use of biosensors. Indeed, the CCD camera can simultaneously capture the bioluminescence produced in each well. Consequently, moving the photosensor from well to well (as with the robotic arm in the system of Cho et al. [42]) is not required.

Yagur-Kroll et al. [43] reported the development of a microfluidic biosensor for assessing general or specific toxicity in which the bioluminescent bacteria are freeze-dried directly onto a disposable microcard made of porous aluminium oxide. This card retains the cells, and its high porosity allows undisturbed access to both the sample and the nutrients.

3 Conclusion

Since the first application of bacterial bioluminescence as a toxicity marker in 1959 [3], many innovations in the domain of biosensors have allowed for the development of robust devices dedicated to environmental monitoring.

The main technological applications of bioluminescent bacterial biosensors were described in this chapter. These biosensors are based on three main strategies for cellular confinement (i.e., free, immobilised, or freeze-dried cells) and induce technological constraints that are specific to their architectures. Nevertheless, several technological advances were proposed for each strategy, always with the objective of the improvement of their use (i.e., simplification, robustness, and reproducibility).

However, a trend seems to be emerging in which the biosensors based on free cells or immobilised cells will not be transferable to industry. Indeed, their complex implementation may inhibit their use in real environmental monitoring applications. Conversely, the trend of using BBBs based on freeze-dried cells, similar to the strategy of the last published system proposed by Yagur-Kroll et al. [43], need to be developed. This strategy greatly simplifies the use of biosensors for everybody (including novices) and allows for the long-term conservation of bioelements without functional alterations.

References

1. Thevenot DR, Toth K, Durst RA, Wilson GS (2001) Electrochemical biosensors: recommended definitions and classification. Biosens Bioelectron 16:121–131
2. Belkin S (2003) Microbial whole-cell sensing systems of environmental pollutants. Curr Opin Microbiol 6:206–212
3. Strehler BL (1959) Some optical properties of luminous bacteria. Arch Biochem Biophys 85:391–408
4. Serat WF, Budinger FE Jr, Mueller PK (1967) Toxicity evaluation of air pollutants by use of luminescent bacteria. Atmospheric Environ 1967(1):21–32
5. Heitzer A, Malachowsky K, Thonnard JE, Bienkowski PR et al (1994) Optical biosensor for environmental on-line monitoring of naphthalene and salicylate bioavailability with an immobilized bioluminescent catabolic reporter bacterium. Appl Env Microbiol 60:1487–1494
6. Iams H, Salzberg B (1935) The secondary emission phototube. Proc Inst Radio Eng 23:55–64
7. Mölders H (1990) Methods for detecting the presence of mercury using microorganisms with mercury-enhanced bioluminescence
8. Bjerketorp J, Hakansson S, Belkin S, Jansson JK (2006) Advances in preservation methods: keeping biosensor microorganisms alive and active. Curr Opin Biotechnol 17:43–49
9. Melamed S, Elad T, Belkin S (2012) Microbial sensor cell arrays. Curr Opin Biotechnol 23:2–8
10. Wang X, Lu X, Chen J (2014) Development of biosensor technologies for analysis of environmental contaminants. Trends Environ Anal Chem 2:25–32
11. Engstrom RW (1980) Photomultiplier handbook. RCA Corp
12. Iams HE, Salzberg B (1935) The secondary emission phototube. Proc IRE 23:55–64
13. Zworykin VK, Morton GA, Malter L (1936) The secondary emission multiplier—a new electronic device. Proc Inst Radio Eng 24:351–375

14. Beijerinck MW (1889) Le Photobacterium luminosum, Bactérie lumineuse de la Mer du Nord. Arch Néerl Sci Exactes Nat 401–27
15. Carmi OA, Stewart GS, Ulitzur S, Kuhn J (1987) Use of bacterial luciferase to establish a promoter probe vehicle capable of nondestructive real-time analysis of gene expression in Bacillus spp. J Bacteriol 169:2165–2170
16. Polyak B, Bassis E, Novodvorets A, Belkin S et al (2001) Bioluminescent whole cell optical fiber sensor to genotoxicants: system optimization. Sens Actuators B Chem 74:18–26
17. Ivask A, Green T, Polyak B, Mor A et al (2007) Fibre-optic bacterial biosensors and their application for the analysis of bioavailable Hg and As in soils and sediments from Aznalcollar mining area in Spain. Biosens Bioelectron 22:1396–1402
18. Hakkila K, Green T, Leskinen P, Ivask A et al (2004) Detection of bioavailable heavy metals in EILATox-Oregon samples using whole-cell luminescent bacterial sensors in suspension or immobilized onto fibre-optic tips. J Appl Toxicol 24:333–342
19. Cheol Gil G, Mitchell RJ, Tai Chang S, Bock Gu M (2000) A biosensor for the detection of gas toxicity using a recombinant bioluminescent bacterium. Biosens Bioelectron 15:23–30
20. Gu MB, Chang ST (2001) Soil biosensor for the detection of PAH toxicity using an immobilized recombinant bacterium and a biosurfactant. Biosens Bioelectron 16:667–674
21. Chang ST, Lee HJ, Gu MB (2004) Enhancement in the sensitivity of an immobilized cell-based soil biosensor for monitoring PAH toxicity. Sens Actuators B Chem 97:272–276
22. Yolcubal I, Piatt JJ, Pierce SA, Brusseau ML et al (2000) Fiber optic detection of in situ lux reporter gene activity in porous media: system design and performance. Anal Chim Acta 422:121–130
23. Valdman E, Gutz IGR (2008) Bioluminescent sensor for naphthalene in air: cell immobilization and evaluation with a dynamic standard atmosphere generator. Sens Actuators B Chem 133:656–663
24. Horry H, Charrier T, Durand MJ, Vrignaud B et al (2007) Technological conception of an optical biosensor with a disposable card for use with bioluminescent bacteria. Sens. Actuators B Chem 122:527–534
25. Mendis S, Kemeny SE, Fossum ER (1994) CMOS active pixel image sensor. IEEE Trans Electron Devices 41:452–453
26. Simpson ML, Sayler GS, Applegate BM, Ripp S et al (1998) Bioluminescent-bioreporter integrated circuits form novel whole-cell biosensors. Trends Biotechnol 16:332–338
27. Lee JH, Mitchell RJ, Kim BC, Cullen DC et al (2005) A cell array biosensor for environmental toxicity analysis. Biosens Bioelectron 21:500–507
28. Sakaguchi T, Morioka Y, Yamasaki M, Iwanaga J et al (2007) Rapid and onsite BOD sensing system using luminous bacterial cells-immobilized chip. Biosens Bioelectron 22:1345–1350
29. Roda A, Cevenini L, Michelini E, Branchini BR (2011) A portable bioluminescence engineered cell-based biosensor for on-site applications. Biosens Bioelectron 26:3647–3653
30. Charrier T, Chapeau C, Bendria L, Picart P et al (2011) A multi-channel bioluminescent bacterial biosensor for the on-line detection of metals and toxicity. Part II: technical development and proof of concept of the biosensor. Anal Bioanal Chem 400:1061–1070
31. Tani H, Maehana K, Kamidate T (2004) Chip-based bioassay using bacterial sensor strains immobilized in three-dimensional microfluidic network. Anal Chem 76:6693–6697
32. Jouanneau S, Durand MJ, Thouand G (2012) Online detection of metals in environmental samples: comparing two concepts of bioluminescent bacterial biosensors. Environ Sci Technol 46:11979–11987
33. Affi M, Solliec C, Legentillomme P, Comiti J et al (2009) Numerical design of a card and related physicochemical phenomena occurring inside agarose-immobilized bacteria: A valuable tool for increasing our knowledge of biosensors. Sens Actuators B Chem 138:310–317
34. Gu MB, Dhurjati PS, Van Dyk TK, LaRossa RA (1996) A miniature bioreactor for sensing toxicity using recombinant bioluminescent escherichia coli cells. Biotechnol Prog 12:393–397
35. Gu MB, Gil GC, Kim JH (1999) A two-stage minibioreactor system for continuous toxicity monitoring. Biosens Bioelectron 14:355–361

36. Gu BM, Gil CG (2001) A multi-channel continuous toxicity monitoring system using recombinant bioluminescent bacteria for classification of toxicity. Biosens Bioelectron 16:661–666
37. Lee JH, Gu MB (2005) An integrated mini biosensor system for continuous water toxicity monitoring. Biosens Bioelectron 20:1744–1749
38. Horry H, Durand M-J, Picart P, Bendriaa L et al (2004) Development of a biosensor for the detection of tributyltin. Environ Toxicol 19:342–345
39. Ikariyama Y, Nishiguchi S, Koyama T, Kobatake E et al (1997) Fiber-optic-based biomonitoring of benzene derivatives by recombinant *E. coli* bearing luciferase gene-fused TOL-plasmid immobilized on the fiber-optic end. Anal Chem 69:2600–2605
40. Rabner A, Belkin S, Rozen R, Shacham Y (2006) Whole-cell luminescence biosensor-based lab-on-chip integrated system for water toxicity analysis. In: Proceedings of SPIE 6112, Microfluidics, BioMEMS, and medical microsystems vol IV, pp 611205–10
41. Choi SH, Gu MB (2002) A portable toxicity biosensor using freeze-dried recombinant bioluminescent bacteria. Biosens Bioelectron 17:433–440
42. Cho J-C, Park K-J, Ihm H-S, Park J-E et al (2004) A novel continuous toxicity test system using a luminously modified freshwater bacterium. Biosens. Bioelectron 20:338–344 (special issue honour Profr. Pierre Coulet)
43. Yagur-Kroll S, Schreuder E, Ingham CJ, Heideman R et al (2015) A miniature porous aluminum oxide-based flow-cell for online water quality monitoring using bacterial sensor cells. Biosens Bioelectron 64:625–632

Part III
Applications of Bioluminescence in Agriculture and Bioprocess

Adv Biochem Eng Biotechnol (2016) 154: 119–146
DOI 10.1007/10_2014_280

Published Online: 14 November 2014

Let There Be Light! Bioluminescent Imaging to Study Bacterial Pathogenesis in Live Animals and Plants

Issmat I. Kassem, Gary A. Splitter, Sally Miller and Gireesh Rajashekara

Abstract Bioluminescence imaging (BLI) of bacteria was primarily designed to permit real-time, sensitive, and noninvasive monitoring of the progression of infection in live animals. Generally, BLI relies on the construction of bacterial strains that possess the *lux* operon. The *lux* operon is composed of a set of genes that encode the luciferase enzyme and its cognate substrate, which interact to produce light—a phenomenon that is referred to as bioluminescence. Bioluminescence emitted by the bacteria can then be detected and imaged within a living host using sensitive charge-coupled device (CCD) cameras. In comparison to traditional host-pathogen studies, BLI offers the opportunity for extended monitoring of infected animals without resorting to euthanasia and extensive tissue processing at each time point. Therefore, BLI can reduce the number of animals required to generate meaningful data, while significantly contributing to the understanding of pathogenesis in the host and, subsequently, the development and evaluation of adequate vaccines and therapeutics. BLI is also useful in characterizing the interactions of pathogens with plants and the para-host environment. In this chapter, we demonstrate the broad application of BLI for studying bacterial pathogens in different niches. Furthermore, we will specifically focus on the use of BLI to characterize the

I.I. Kassem · G. Rajashekara (✉)
Ohio Agricultural Research and Development Center,
Department of Veterinary Preventive Medicine, The Ohio State University,
Wooster, OH 44691, USA
e-mail: rajashekara.2@osu.edu

G.A. Splitter
Department of Pathobiological Sciences,
University of Wisconsin-Madison School of Veterinary Medicine, Madison
WI, USA

S. Miller
Department of Plant Pathology, Ohio Agricultural Research and Development Center,
The Ohio State University, Wooster, OH 44691, USA

following: (1) the pathogenesis of *Brucella melitensis* in mice (animal host), and (2) the progression of infection of *Clavibacter michiganensis* subsp. *michiganensis* in tomatoes (plant host). These studies will provide an overview of the wide potential of BLI and its role in enhancing the study of unique—and sometimes difficult-to-characterize—bacterial pathogens.

Keywords Bioluminescence imaging · *Lux* operon · *Brucella* · *Clavibacter* · Pathogenesis · Animal model · Plant infections

Contents

1 A Beacon of Light: Origins and Applications of Bioluminescence

It has been long known that certain species of bacteria and other organisms can be naturally bioluminescent, which is generally defined as the ability to emit light via intrinsic or symbiotic biochemical reactions [76, 77]. Although more than 700 species have systems for bioluminescence, this phenomenon is more widely encountered in marine organisms, including planktonic bacteria [77]. The spread of bioluminescence ability across many species strongly suggests that this phenomenon might pose certain evolutionary advantages. These may include direct benefits, such as the attraction of suitable mates, luring prey, camouflage, and within species communication, among others. They also collectively contribute to the appeal of establishing a symbiotic relationship between higher organisms, such as squid and angler fish, and bioluminescent bacteria [76, 77]. The latter usually occur as intestinal flora or are contained in specialized organs that play a role in the regulation of light production [76]. Therefore, the ability for bioluminescence is selected and coveted in certain niches, promoting the survival of the light-emitting symbionts. This is further facilitated by the assumption that the ability to generate bioluminescence in bacteria, while obviously energy consuming, might not be necessarily genetically overtaxing. For example, *Photorhabdus luminescens*, a naturally bioluminescent bacterium and a pathogen of nematodes, possesses a *lux* operon, which encodes for one of the best-studied bioluminescence systems in bacteria [23, 32]. This operon includes five genes (*luxCDABE*) that are deemed to be essential genetic requirements for generating light. Specifically, *luxAB* encodes the α-subunit (40 kDa) and β-subunit (35 kDa) of the luciferase, a heterodimeric oxidative enzyme that serves as a catalyst for the light production reaction [23, 76]. The *luxCD* and *luxE* (*luxC* encodes the acyl-reductase, *luxD* encodes an acyl-transferase, and *luxE* encodes the protein required for acyl-protein synthetase) are involved in the production of a fatty acid reductase complex that facilitates the biosynthesis of a long-chain fatty aldehyde, luciferin [23, 32]. In the presence of oxygen, lucifern (RCHO) and a reduced form of a flavin mononucleotide ($FMNH_2$) are oxidized by luciferase, leading to the production of blue/green light, or bioluminescence [16, 47]:

$$FMNH_2 + RCHO + O_2 \overset{\text{Luciferase}}{\rightarrow} FMN + RCOOH + H_2O + \text{light}$$

Luciferase systems have been used in reporter gene assays, insertional mutagenesis studies, bacteria detection in a variety of matrices, including food, sensing environmental contaminants, tumor imaging, gene expression, and imaging of bacterial infections in live animals [3, 76]. These applications were facilitated by unique intrinsic properties. For instance, luciferases possess significantly shorter half-lives than fluorescent proteins, such as the green fluorescent protein (GFP) [2, 13]. Notably, the half-life of the luciferase of *P. luminescens* can be as low as 46 min after introducing destabilization signals into the enzyme [2]. Furthermore, fluorescent proteins such as the GFP undergo postranslational maturation, which can require several hours [73, 75]. Fluorescent proteins also undergo photobleaching and are associated with a high autofluorescence background inherent to biological samples, while the energetics of luciferase systems allow for enhanced tissue penetration in live animals [60, 73, 76]. Therefore, luciferase systems allow for superior, improved dynamic assessment of expression and are generally considered to be more efficient and less labor-intensive compared to other systems. Alternatively, certain applications of bioluminescence, such as imaging bacterial infections in live animals, may be complicated by the properties of light propagation through tissues that reduce the intensity of light due to scattering and photon diffusion [3].

To bypass these limitations, bioluminescence imaging (BLI) in animal models may require sophisticated equipment to process light signals and capture images. Furthermore, luciferase-dependent light production requires oxygen, which may be a disadvantage in severely oxygen-limited environments and when studying strictly anaerobic bacteria [3]. However, strategies exist to bypass the latter limitation, such as exposing samples to oxygen before quantifying bioluminescence, while the system can be used inversely to assess the development of an anaerobic environment [3]. Also, in studies concerning bioluminescent bacteria, the detection limit may be a limiting factor. In live animals, 10^3 and 10^5 bioluminescent bacteria can be detected, which is influenced by the properties of the infected organ and pigmentation [30, 55, 78]. The detection limit also depends on the sensitivity of the instrumentation/imaging system used to quantify the light signal [3]. However, as technology advances, it is predicted that the detection limit will become more sensitive. Regardless, because luciferase-dependent bioluminescence requires both adenosine triphosphate and oxygen, only living organisms produce light; coupled with the short half-life of luciferase, this reduces errors in the assessment of colonization and/or growth dynamics of a bioluminescent pathogen. Certainly, this was successfully demonstrated by engineering bioluminescence into many important pathogens, such as *Escherichia coli* O157:H7, *Staphylococcus aureus*, and *Pseudomonas aeruginosa* [24, 58, 71].

Although the regulation of bioluminescence in certain bacteria is still not fully understood, advances in characterizing the *lux* operon and cognate enzymes along with the relative simplicity of the minimum genetic requirements for light production highlight the vast biotechnological potential associated with harnessing the bioluminescence phenomenon. However, while BLI of live animals has been receiving considerable attention, the use of this technology in studying plant—bacterial interactions is still evolving. To demonstrate some of the applications associated with BLI, this chapter focuses on the use of bioluminescence to study bacterial pathogens in both a live animal model and a plant system. Therefore, we partition this chapter into two major sections dedicated to detailed examples of the application of BLI in each of the aforementioned hosts.

2 The Light Within: Bioluminescence as a Tool to Study Host–Pathogen Interactions in Live Animal Models

Live animal models are generally accepted as crucial for studying the interactions of a bacterial pathogen with a host, including assessment of disease symptoms, host immunological responses, and bacterial pathogenesis, as well as the efficacy of new medications and interventions, such as the use of antibiotic treatments and vaccine development. Therefore, animal models play an integral part in the advancement of the fields of microbiology and infectious diseases. Working with certain animal models can be labor-intensive and expensive, while ethical and humane concerns are always of paramount importance when experimenting on sentient organisms. Subsequently, live animal model studies are always subjected to reduction and refinement to reduce the unnecessary suffering of animals and maximize the yield and statistical significance of generated data.

Historically, investigations of infections in animal models have been compounded by "endpoint" measurements, which occur after euthanizing animals to harvest selected tissues and quantify associated bacterial colony forming units and/or perform histopathological or immunological analysis [11]. This approach is temporally limited, allowing neither monitoring of the progression of infection throughout the entire life cycle of the pathogen, nor a proper illustration of the role of the host's immunity in maintaining the infection. This is especially true in the cases where pathogens can cause persistent, as well as acute infections [11]. Furthermore, endpoint experiments mandate the use of numerous animals to generate reliable data and account for variability between individual hosts. The localization of the pathogen to previously unexpected host organs/tissues and cognate interactions can go unnoticed due to the selective sampling of tissues [11]. For example, using a sensitive imaging method (discussed later), it was found that *Brucella melitensis* can replicate in the salivary glands of IRF-1$^{-/-}$and wild-type C57BL/6 mice, which was previously an unknown tissue preference [55]. Similarly, *Listeria monocytogenes*, previously believed to be mainly an intracellular pathogen, was found to replicate extracellularly and asymptomatically in the mouse gallbladder, which provided new insight into the location of this pathogen in potential human carriers [34]. Furthermore, in cases where the affected organ is not known, end-point sampling must be widely inclusive of different tissues, which can be resource and labor intensive. Taken together, the aforementioned observations strongly indicate the need for advanced methods that transcend the limitations of endpoint studies of bacterial infections in live animal models.

A method of interest is BLI, which can facilitate noninvasive, sensitive, and real-time monitoring of bacteria in disparate niches, including hosts and the para-host environment [11, 33]. BLI relies on bacteria that have been genetically engineered to carry the *lux* operon and emit light [16]. Subsequently, the introduction of the bioluminescent pathogen into the host can be monitored using a highly sensitive charged-coupled device (CCD) camera, which permits following infection in real time in live animals. Therefore, BLI allows assessment of the following: (1) temporal and spatial dissemination pattern of the bacterial pathogen within the host, (2) different dissemination patterns between virulent and nonvirulent or mutant strains, (3) location of the bacteria in chronically infected animals, (4) level of persistence and duration of colonization in an animal, (5) progression of infection in different hosts, including immunologically compromised individuals, and (6) impact of interventions such as antibiotic treatment and vaccines on the bacterial colonization of the host. Subsequently, BLI provides crucial insight into the mechanisms of bacterial pathogenesis and host response, improving our understanding of host–pathogen interactions and allowing a thorough and noninvasive assessment of infection progression and persistence, as well as designing efficient interventions using live animals.

These virtues of BLI have been successfully used to study the pathogenesis of a plethora of microorganisms (see [3] for a list of microorganisms used for noninvasive imaging). However, here we highlight the use of BLI to study the pathogenesis of *B. melitensis*, for the following reasons: (1) *Brucella* infections are generally difficult to study, (2) BLI provided crucial and novel insights into the pathogenesis of this bacterium in a live animal model, and (3) the studies using bioluminescent *Brucella* provide an ideal summary for the aforementioned virtues of BLI.

3 Use of Bioluminescence Imaging to Evaluate the Pathogenesis of *Brucella melitensis* in a Live Animal Model

3.1 Brucella *and Brucellosis in Brief*

Brucella spp. are Gram-negative facultative intracellular bacteria that can cause disease in humans and animals. In animals, *Brucella* can cause orchitis, abortions, or the birth of severely debilitated offspring; however, the disease has few other clinical manifestations. These pathogens can be transmitted via infected milk, placenta, or aerosol, while they can survive and persist intracellularly in macrophages and placental cells. *Brucella* infections in humans are chronic and are acquired by direct contact, including inhalation, or consumption of contaminated animal products [25].

Human brucellosis is prevalent in underdeveloped countries and is associated with several acute symptoms, including undulating fever, malaise, sweats, arthralgia, lower back pain, splenomegaly, osteoarticular involvement, cervical lymphadenitis, hepatomegaly, genitourinary involvement, and cholecystitis [31, 81]. The severity of brucellosis and the properties of the etiologic agent, mainly aerosolization and infection of different hosts, have raised fears of using brucellae in bioterrorism. It has been predicted that the aerosolization of brucellae in a suburb of a major city can result in severe health and economic repercussions, including an estimated cost of $477.7 million per 100,000 people exposed [39]. Therefore, brucellosis is a significant disease that is also considered to be a contributor to the perpetuation of poverty [46].

Despite the importance of the disease, until recently little was known about the pathogenesis of *B. melitensis*, the species that is frequently associated with brucellosis. Although brucellosis can be studied in mouse models, much remains unknown regarding the dissemination and tissue localization of *B. melitensis*. However, the recent application of BLI to monitor the pathogen in real time in a living animal has provided answers to important questions regarding host–pathogen interactions that were either difficult to address or could not be addressed using classical experimental approaches. Specifically, BLI of *Brucella* infections in live animals can be used to investigate the following important questions:

1. What is the dissemination pattern of virulent *B. melitensis* and how does the dissemination pattern differ with nonvirulent or mutant bacteria?
2. Where does *B. melitensis* remain in a chronically infected animal?
3. What is the level of persistence, duration, and location of live attenuated *Brucella* vaccines in an animal?
4. Can antibiotic treatment be monitored to distinguish recovery or relapse?

Using BLI to address these questions has provided insights into the mechanisms of *Brucella* pathogenesis and improved our understanding of vaccine development against this hazardous pathogen. In the next section, we discuss the construction of bioluminescent *B. melitensis* and demonstrate the virtues of deploying BLI for studying host–pathogen interaction events in living mice in real time.

3.2 Construction and Selection of Bioluminescent Brucella *Strains*

To construct bioluminescent strains (Fig. 1), electrocompetent *B. melitensis* were prepared using pure isolates that were grown in 10 ml *Brucella* broth to achieve an OD_{600} (optical density at 600 nm) of 0.5–0.7. The cultures were chilled on ice for 10 min, and the pellets were collected by centrifugation (5 min at 5,000 g). This was followed by washing the cells twice with sterile deionized water, and the pellets were then resuspended in water to 1/50th of the original volume. 50 μl of the competent cells were mixed with 4 μl of 7-kb linear DNA fragments containing the EZ::TN-*lux*. The DNA fragments were produced by digesting the 10-kb plasmid (pUWGR4), which was constructed using a promoterless *luxCDABE* operon that was inserted into the transposon vector EZ::TN pMod-3 ⟨R6Kγori/MCS⟩ [55]. The latter was modified to carry a kanamycin resistance gene (Kan^R) marker [55]. It should be noted that the luciferase of *P. luminescence* was selected for the construction of bioluminescent strains because of its thermostability and optimum activity at mammalian body temperatures [23, 45, 69].

The cell-transposome mixture was subjected to electroportation (2.5 kV, 25 μF, and 200 Ω) in cuvettes that were prechilled on ice. Following electroporation, the cells were suspended in Super Optimal Broth containing glucose and incubated, shaking for 7 h. The suspension was spread onto *Brucella* agar plates supplemented with kanamycin (50 μg/ml), which were incubated at 37 °C under 5 % CO_2 for 5–7 days. The resulting colonies were screened for bioluminescence using the In Vivo Imaging System (IVIS; Xenogen), a dark imaging chamber supplied with a highly sensitive CCD camera.

Bioluminescent colonies were streak purified and grown in *Brucella* broth containing kanamycin to quantify the amount of bioluminescence using the IVIS machine. Strains with the highest bioluminescence were selected for mice infection studies [54, 55]. It is important to note that this approach results in the insertion of the promoterless *lux* operon in the genome of *B. melitensis*. As such, the intensity of

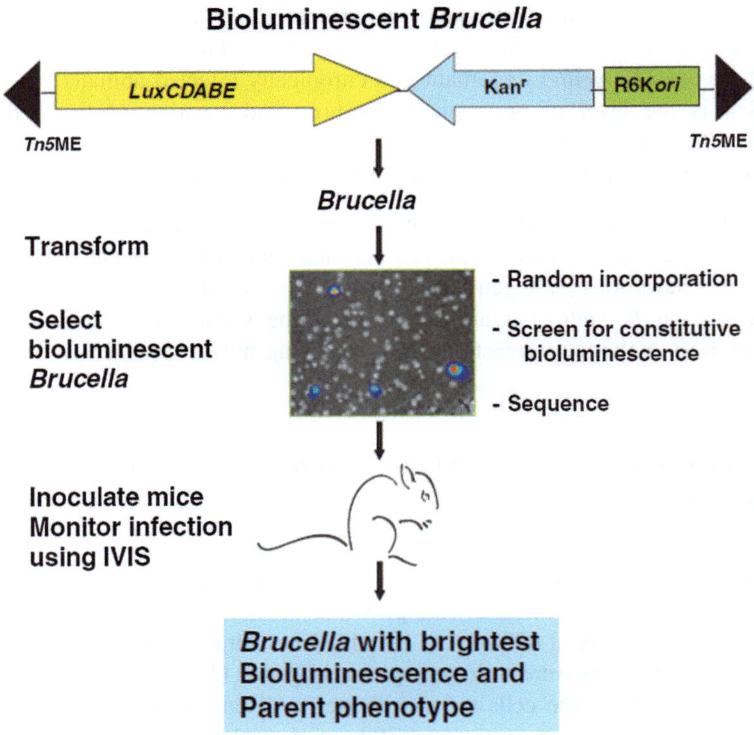

Fig. 1 Schematic representation for the construction and selection of bioluminescent *Brucella* strains used to infect mice. Bioluminescent strains were obtained using a modified EZ::TN transposon containing a promoterless *lux* operon that was inserted into the genome of *Brucella*. The strains were then isolated and screened for bioluminescence using the IVIS, an in vivo imaging system (Xenogen Corporation, USA). Selected strains that constitutively expressed a strong bioluminescent signal were used to infect mice [56]

bioluminescence will be regulated by an indigenous bacterial promoter, which can vary in the type (constitutive vs. inducible) and strength of cognate expression. Therefore, screening of the *lux*-transformed colonies to select for the strongest and most stable bioluminescence is crucial for downstream applications. It is also important to map the insertion site of the *lux* operon in the genome of the bacterium and determine that growth phenotypes, physiology, and virulence of the resulting bioluminescent strains are not impaired compared to the parental strain. For this purpose, the infection properties of bioluminescent and wild-type strains were compared in mice.

It was found that, like the *B. melitensis* 16M (wild-type) strain, the strongly bioluminescent strain, GR023 [1×10^7 colony-forming units (CFU) injected intraperitoneally], resulted in the death of the immunocompromised IRF-1$^{-/-}$ mice within 10 days [54, 55]. This was important because the lifespan of these mice is affected by the virulence and the density of the *Brucella* strain used in infection [41, 42].

For example, using a higher inoculum of a virulent *Brucella* strain for infection (1×10^6 CFU vs. 5×10^3 CFU) reduced the survival of these mice to 21 days as compared to 28 days [55]. Subsequently, the IRF-$1^{-/-}$ mice are adequate for comparing the virulence of the *Brucella* strains [41, 42]. Furthermore, CFU of GR023 quantified from liver and spleen samples of the moribund mice were comparable to those of 16M and the strains resulted in similar histopathological changes in these tissues [55]. Therefore, the virulence of the bioluminescent strain (GR023) selected for animal studies was not compromised.

3.3 Knowledge Garnered From the Use of Bioluminescence to Study Brucella *in a Live Animal Model*

Selected bioluminescent *B. melitensis* was introduced into mice, including IRF-$1^{-/-}$, BALB/c, and C57BL/6, which are the most common mouse strains used to study the pathogenesis of *Brucella*. The infected mice were then monitored using the IVIS system to detect and quantify bioluminescent signals. To confirm that the localization of the bioluminescent signal was actually indicative of the occurrence of the bacterial strains, organs were harvested as appropriate to quantify bacterial CFU numbers. The isolated bacteria were also imaged to confirm that they were bioluminescent. After confirmation, it was apparent that the CFU counts correlated strongly with photon counts in infected mice. The BLI detection limit in the mice ranged between 100 and 1,000 bacteria and was dependent on the location of the area of interest; that is, the detection limit was optimal near the surface and declined in deeper organs [54–56].

3.3.1 BLI for Monitoring the Localization of Bioluminescent *B. melitensis* in Mice

Using the conditions and parameters listed above, bioluminescent *B. melitensis* was readily detected in the inguinal lymph nodes, spleen, liver, testes, and the osteoarticular and submandibular regions of infected mice (Fig. 2), suggesting a systemic spread of the pathogen [55]. These observations indicated that the localization of the pathogen in mice organs was similar to that reported in humans. For example, previous reports showed that *B. melitensis* is frequently detected in the testes of infected humans and domestic animals [43, 44], where it may be associated with orchitis.

B. melitensis was localized at early stages to the oral cavity, where it occurred in salivary glands [55]. Bioluminescent signals were detected in the oral cavity of mice, regardless of whether the pathogen was introduced via intranasal, oral, or even intraperitoneal infections (Fig. 3) (unpublished data). Because oral ingestion is an important route of exposure in humans and animals, early localization to the oral niche may constitute a precursor for adaptation to the host, instigating cellular events such as the expression of genes that promote survival and virulence [55].

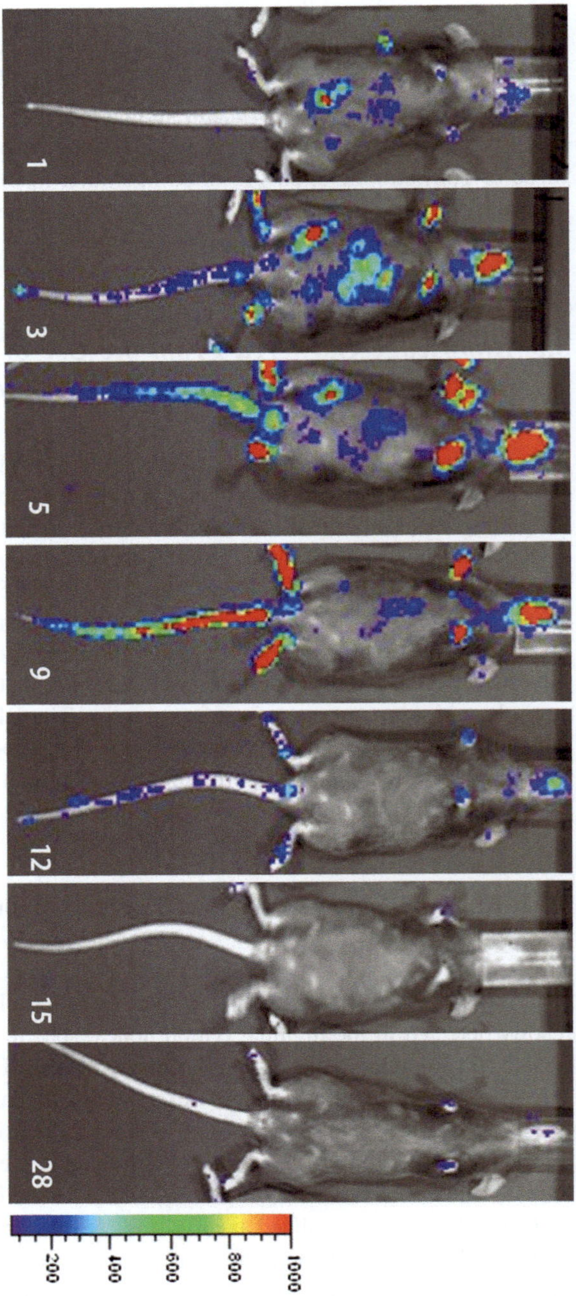

Fig. 2 Localization of bioluminescent virulent *B. melitensis* strain GR023 in C57BL/6 mice. The *B. melitensis* strain GR023 (1×10^7 CFU) was introduced into the lower peritoneal cavity and infection was monitored for 28 days. A large number of bacteria were present in the oral cavity at days 3–14 postinfection (DPI). The numbers in the lower right of each panel indicate the DPI. The rainbow scale represents approximate photon counts [55]

BLI also revealed that mice infected in the lower peritoneal cavity revealed bioluminescent signals in the tail, coccygeal vertebrae, and joints of the feet (Fig. 4) [55; unpublished data]. Furthermore, bioluminescent bacteria were detected in the

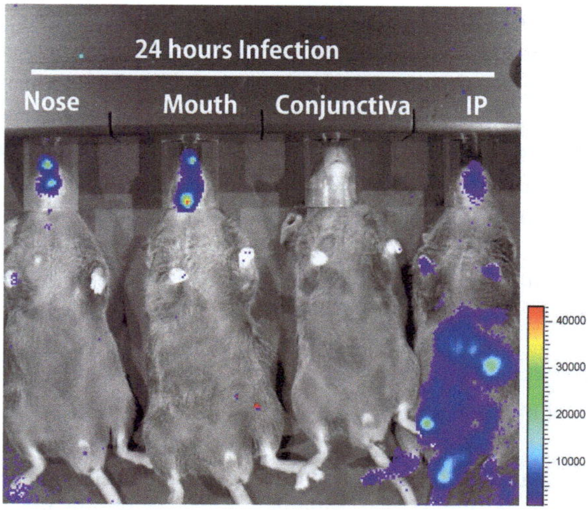

Fig. 3 BLI at 24 h of IRF-$1^{-/-}$ mice infected with *B. melitensis* GR023 (1×10^7 CFU) via the nose, mouth, conjunctiva or intraperitoneal cavity. The rainbow scale represents approximate photon counts (unpublished data)

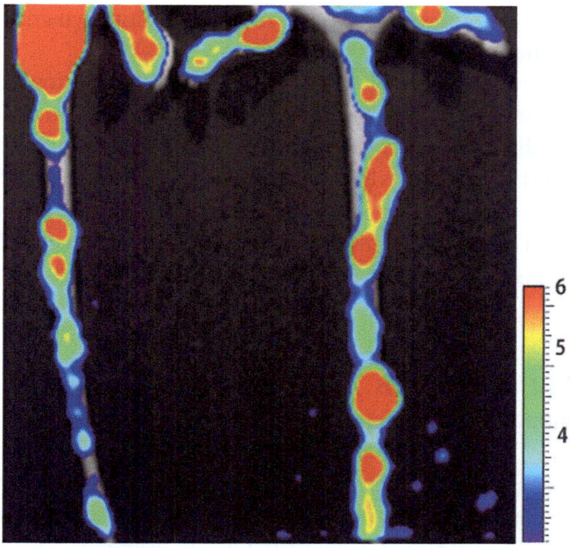

Fig. 4 BLI of the tail and feet of BALB/c mice infected with bioluminescent virulent *B. melitensis* strain GR023 (1×10^6 CFU). The bioluminescent strain GR023 was introduced at the lower peritoneal cavity. The bioluminescent signal is clearly localized to the segments of coccygeal vertebrae as well as the joints of the feet. The rainbow scale represents approximate photon counts [55; unpublished data]

region of the heart of similarly infected mice (unpublished data). These observations are notable because chronic brucellosis in humans is associated with multiple symptoms including orchitis, osteoarthritis, spondylitis, and endocarditis [12, 81].

In humans, *Brucella* in the blood is removed by cells in the liver, spleen, and bone marrow, whereas chronic brucellosis rarely results in suppurative disease of the liver and spleen [74]. In C57BL/6 mice, bioluminescence was detected in the liver and spleen, the primary organs for *Brucella* replication. While the signal persisted in the liver of these mice up to 21 days postinfection, bioluminescence was not detected in the spleen at day 14. However, processing spleen samples at day 14 yielded 4.8×10^3 *Brucella* CFU. This was explained by the presence of a capsular sheath that might interfere with light detection, rendering the spleen less sensitive to BLI [55].

Furthermore, immunocompromised IRF-1$^{-/-}$mice that were inoculated intraperitoneally with 1×10^7 CFU of bioluminescent *Brucella* died at 7–10 days postinfection. Usually at day 7 postinfection, bioluminescence was detected throughout these mice, which was followed by death within the following 24 h. Liver and spleen samples from the moribund mice were infected with greater than 10^9 *Brucella* CFU.

Taken together, these observations suggest the following:

1. BLI of mice infected with bioluminescent *B. melitensis* represents an adequate model for assessing both acute and chronic *Brucella* infections.
2. *Brucella* infection progression and tissue localization dynamics can be monitored in real time in the host.
3. The ability to image the whole body of the same infected mouse across time is crucial to assess the protean nature of *Brucella* infections.
4. BLI revealed that *Brucella* can replicate in the salivary glands of IRF-1$^{-/-}$ and wild-type C57BL/6 mice, which was previously unknown and may possibly facilitate initial amplification and subsequent spread of the pathogen.
5. Localization of *Brucella* to the joints at later stages, beyond 28 days postinfection, may provide a model to understand chronic brucellosis in humans.

3.3.2 Use of BLI to Identify *B. melitensis* Virulence Factors and Evaluate Vaccine Candidates

During the screening of different bioluminescent *Brucella* that were generated by transposon insertion, in one strain, GR019, the *lux* operon was inserted in *virB4*, which encodes an ATPase [55]. The VirB type IV secretory system (T4SS) contributes to the survival of *Brucella* in macrophages and plays a role in the establishment and maintenance of persistent infections in mice [14, 22, 35, 50, 66]. Subsequently, the use of GR019 in mice infection studies would generate further important insights regarding the role of VirB in the virulence of *Brucella*. This was also feasible because GR019 exhibited strong bioluminescence in vivo [55].

BLI of IRF-$1^{-/-}$ mice infected with GR019 demonstrated that the impairment of *virB4* resulted in less extensive signals, but the initial systemic spread of the pathogen was comparable to that of other strains. Specifically, GR019 was detected in several locations, including the livers, spleens, and testes by day 3 [54, 55]. However, daily monitoring showed that by day 4, the bioluminescent signal of GR019 in the host decreased in most tissues. Quantification of *Brucella* CFU in livers and spleens corroborated the BLI data, while these tissues did not exhibit detectable histologic changes [54, 55]. Although no signal associated with the *virB4* mutant was detected in the tail at any time, a weak bioluminescence was observed, mainly in the feet and submandibular area by day 24 [55]. As compared to the bioluminescence trends in IRF-$1^{-/-}$ mice, BLI showed that the spread of the *virB4* mutant in C57BL/6 mice was also systemic. However, the bacterial strain was cleared earlier, with almost no bioluminescence detected by day 7 postinfection. Therefore, BLI showed that the *virB4* mutant could still disseminate and persist in the host for days prior to immune elimination. Taken together, data collected using BLI confirmed the importance the VirB type IV secretory system in the localization and persistence of *Brucella* [55, 56].

As importantly, transposon dependent *lux* insertion and subsequent BLI imaging of the bioluminescent variant strains in IRF-$1^{-/-}$ mice might constitute an efficient system for the identification of bacterial genes involved in virulence and persistence in the host. This conclusion was further confirmed in another study that deployed bioluminescent *Brucella* variant strains to study the role of a TIR domain containing protein, TcpB (also termed Btp1), of *B. melitensis* during infection in mice [53]. BLI revealed that TcpB was required for the initial growth and spread of *B. melitensis* in vivo [53]. In conclusion, BLI of bioluminescent *B. melitensis* mutants facilitates the identification in real time of genes contributing to in vivo dissemination, persistence, and localization of this pathogen.

A subsequent and readily apparent outcome of the use of BLI is the identification of *Brucella* strains that were attenuated in mice. This is of peremptory importance because attenuated strains constitute potential vaccine candidates. For this purpose, Rajashekara et al. [54] identified and evaluated bioluminescent strains with mutation in the *virB4, galE,* and *BMEI1090-BMEI1091*, respectively. In addition to its properties listed above, the *virB4* mutant was attenuated in macrophage cultures, while the bioluminescent strains GR024 (*galE* mutant) and the GR026 (*BMEI1090-BMEI1091* mutant) were not attenuated in macrophages, but their dissemination in vivo was highly limited [54].

Interestingly, BLI confirmed the results of Rajashekara et al. [54], demonstrating that prior vaccination of IRF-$1^{-/-}$ mice with these mutants limited the dissemination of the bioluminescent virulent GR023, possibly enhancing host immunity and ensuring the survival of these mice that typically die from virulent *Brucella* infections (Fig. 5). Taken together, these findings confirm that BLI allows the assessment of host–pathogen interactions to develop and evaluate potential vaccines.

Also of interest is the possibility for the construction of bioluminescent variants of current vaccine strains, such as *B. melitensis* Rev. 1, and *Brucella abortus* RB51

Fig. 5 BLI of IRF-1$^{-/-}$ mice that were vaccinated and challenged with the virulent *B. melitensis* strain GR023. The mice were immunized with an attenuated mutant strain of *B. melitensis*. After 6 weeks, the mice were then challenged with the bioluminescent GR023. Note the minimal dissemination of virulent bacteria following challenge. Numbers under each image represent the day of infection after the challenge. The rainbow scale represents the approximate photon counts [54]

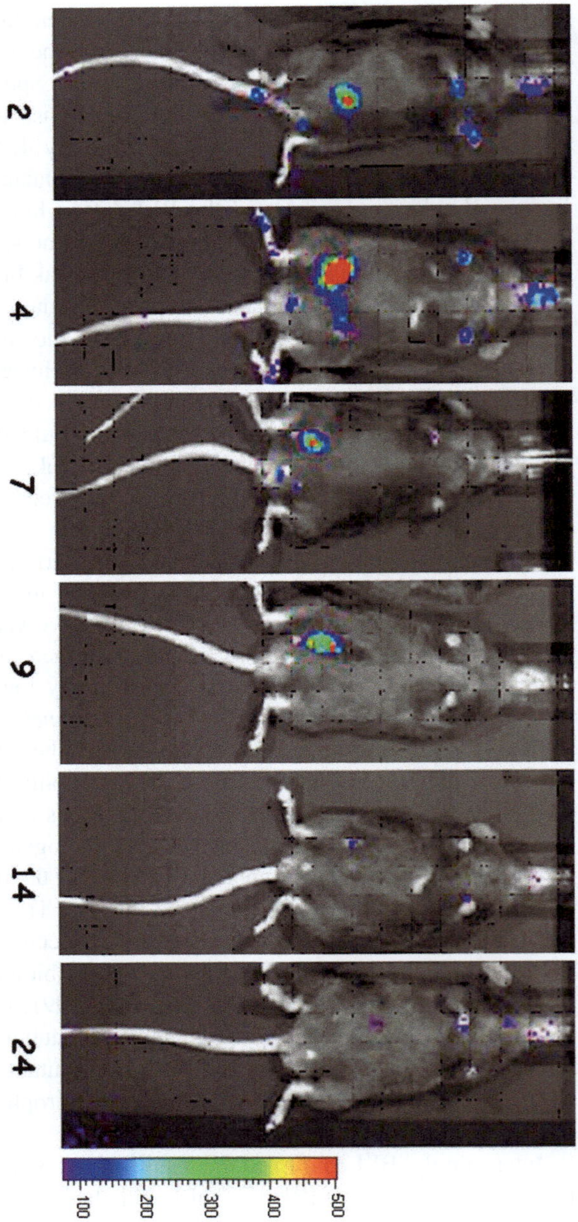

and strain 19. This would provide useful information regarding how current vaccines persist and disseminate in the mouse as benchmarks for alternative vaccine development. The evaluation of vaccine dissemination, persistence in vivo, and cognate levels of immunity pose a question that has been poorly addressed in *Brucella* vaccine development. This is of paramount relevance because highly attenuated *Brucella* vaccines that are rapidly cleared fail to provide protection, whereas *Brucella* vaccines that are still virulent and persist in the host appear to provide higher levels of protection.

3.3.3 Use of BLI to Evaluate the Role of Erythritol in *Brucella* Infections

Erythritol, a four-carbon sugar, is produced in the reproductive tract by several domestic ruminant animals, including goats, cows, and pigs [18]. The sugar is preferentially used by *Brucella* spp. and it has been used to explain the localization and accumulation of *Brucella* in tissues [67]. Specifically, the presence of the sugar in the third trimester of pregnancy is associated with high localization (10^{13} bacteria/gram of tissue) of *Brucella* to the placenta [1], leading to abortion [59].

Interestingly, using the bioluminescent virulent *B. melitensis* strain GR023, it was observed that *B. melitensis* localizes to sites within the mouse where erythritol was artificially introduced [52]. This was confirmed by also introducing glucose as a control, which did not result in preferential localization of the pathogen. The reasons and mechanisms for the localization of *Brucella* to sites containing erythritol are not well characterized, but they may include preferentially active processes in the bacterium (i.e. a chemotactic system to erythritol), placental neovascularization, or an increased level of replication upon reaching the site of erythritol. However, microarray analysis of bacteria grown either with or without erythritol showed the upregulation of two major virulence pathways, namely the type IV VirB secretion system and flagellar proteins, suggesting a role for erythritol in facilitating *Brucella* virulence [52]. Further analysis revealed that erythritol impacted the bacterium intracellularly, because *B. melitensis* was unable to replicate within macrophages in the presence of the sugar [52]. However, this inhibition appeared to encourage the bacteria to replicate extracellularly, which was an unexpected insight into the pathogenesis of *Brucella.*

Taken together, these findings highlight possible immediate advantages associated with using BLI for the following: (1) to design a model system to study the pathogenesis of *Brucella* abortion, and (2) to identify host factors that may influence *Brucella* localization and infections. The latter can also be used for intervention and the design of alternative therapeutics.

3.3.4 Summary of BLI Analysis of Bioluminescent *B. melitensis* in Mice

The advantages of using BLI to study *Brucella* in mice included the following:

1. Monitoring infections in the intact animal, which provides a more complete evaluation of dissemination, persistence, and localization of *Brucella* compared to conventional methods. For example, previously mutant bacteria were typically assessed by CFU quantification from the spleen or liver.

2. Tracing the localization and tissue preference during acute and chronic infections. Furthermore, the BLI approach accounts for the protean nature of *Brucella* (i.e., osetoarticular, cardiac, or testicular involvement).
3. Monitoring the kinetics of dissemination and localization within the host using the same mouse. This allowed temporal assessment of infection, which accounted also for host-dependent variability. The approach reduced the number of mice needed for statistical analysis as opposed to using a separate mouse for each time point.
4. Identification of attenuated *Brucella* strains that can be used in vaccine development.
5. Identification of *Brucella* genes involved in pathogenesis.
6. Identification of host factors involved in *Brucella* localization and virulence.

4 Using Bioluminescence Imaging to Study the Pathogenesis of *Clavibacter michiganensis* subsp. *michiganensis*

4.1 BLI and the Study of Bacterial Pathogen–Plant Interactions

Challenging plants with a pathogen is not constrained by the same ethical limitations encountered during the study of infections in animal models. Although BLI offers the advantage of reducing the number and, subsequently, the suffering of infected animals, the latter is not a factor in the experimentation on plant–pathogen interactions. However, other advantages, such as reducing labor by processing fewer plant samples, better temporal analysis of localization, limiting host variability, identification of pathogen and host factors that contribute to virulence and disease development, and identification of plant traits that result in resistance to infection, also apply to BLI of plant–pathogen interactions. BLI can also be useful in investigating the internalization and transmission of the pathogen between plants as well as tracing the source of the infection, which may include contaminated grafts, seeds, and soils.

The *luxCDABE* genes have been used on multiple occasions to generate bioluminescent beneficial and pathogenic bacteria for plant studies, which have included the following:

1. The assessment of differential gene expression of bacteria during interaction with the plant
2. Identification of genetic loci involved in pathogenesis
3. Monitoring internalization and localization of bacteria in plant tissues
4. Analysis of plant susceptibility (symptom development)
5. Tracing the source and spread of contamination [5, 8, 9, 15, 20, 26, 37, 51, 63, 64, 79, 80].

For example, bioluminescent strains were used to examine the root colonization of maize and lettuce with *Rhizobium leguminosarum* biovar *phaseoli* [5]. Furthermore, bioluminescent strains were constructed to study the movement and distribution of phytopathogenic *Xanthomonas campestris* pathovars and *Pseudomonas* spp. in host plants [8, 26, 51, 63]. Fukui et al. [26] used a bioluminescent strain of *X. campestris* pv. *dieffenbachiae* to study the relationship between symptom development and the site of infection in the leaves of anthurium.

Taken together, these studies confirm the breadth of possible applications as well as the advantages of using bioluminescence to study plant–bacterial interactions. In the next section, we further demonstrate the usefulness of BLI by discussing insights garnered from using this technology to analyze an economically significant and notoriously difficult-to-control plant pathogen, *C. michiganensis* subsp. *michiganensis*.

4.2 **Clavibacter michiganensis** *subsp.* **michiganensis** *in Brief*

Clavibacter michiganensis subsp. *michiganensis* (*Cmm*) is a rod-shaped, Gram-positive, aerobic actinomycete that causes bacterial canker in tomato plants [27, 68]. The disease is characterized by impaired water transport that results in plant wilting, stunting, and death [27, 36]. The "window of vulnerability" during which tomato plants are highly susceptible to *Cmm* has been described as the time from transplanting to first fruit set [62]. Plants infected after that time develop symptoms but the yield is not reduced, despite the fact that bacterial titers in the plant may be high. Plants infected with *Cmm* late in the production cycle may also remain asymptomatic, which contributes to the production of contaminated seeds [21, 57]. The latter is an important route for disseminating the pathogen over long distances.

Cmm colonizes the lumen of xylem vessels [4, 6, 48] through which it reaches and colonizes seeds; it may also colonize seeds through lesions on fruits [70]. Seedlings originating from infected seeds are often symptomless; however, they may harbor latent infections or carry the pathogen epiphytically [28, 65, 72]. Studies showed that seedlings grown from artificially and naturally infested seeds harbored *Cmm*, ranging from 10^4 to 10^8 CFU/seedling and 10^3 to 10^4 CFU/g shoots, respectively [65, 72]. Therefore, infected seeds or transplants constitute a major source of *Cmm* introduction and subsequent outbreaks of bacterial canker in tomato production [7, 21, 28, 68]. *Cmm* is spread mechanically from plant to plant by numerous routes, including contaminated tools during grafting operations and crop work (tying, pruning, etc.), transfer of guttation fluids from infected plants [61], and water.

Bacterial canker was first reported in 1909 in greenhouses in the state of Michigan (USA); however, since then the disease has been observed in tomato production areas around the world [19]. The aforementioned properties of *Cmm* infections suggest that cognate diseases can be devastating for the crop, resulting in severe economic losses [49, 68]. Bacterial canker is widely considered the most important disease of greenhouse-produced tomatoes, and *Cmm* is a quarantine pathogen in Europe. Bacterial canker causes up to 80 % yield loss due to the wilting and death of plants, lesions on fruits, and reduced photosynthetic capacity, fruit set, and size [49, 68, 80].

The control of *Cmm* is challenging. Even a low *Cmm* transmission rate (0.01 %) from seed to seedling can cause a disease epidemic under favorable conditions [7], while highly canker-resistant commercial tomato cultivars and adequate bactericides are not currently available [36]. Therefore, in field tomatoes, controlling bacterial canker largely relies on producing pathogen-free seeds and transplants, although these tactics are not fully deployed and outbreaks continue to occur [36, 80]. Once introduced in greenhouse tomato production systems, bacterial canker can only be managed by removal of diseased and asymptomatic neighboring plants, quarantine of areas from which diseased plants were removed, and strict and extensive sanitation practices.

Mechanisms of virulence and pathogenicity of *Cmm* in tomato have been reviewed recently [17]. Briefly, *Cmm* carries two circular plasmids, pCM1 and pCM2, both of which are required for full symptom development. These plasmids carry the genes *celA* and *pat-1*, which encode an endo-ß-1,4-glucanase and a serine protease, respectively. Strains cured of both plasmids replicate in tomato plants as endophytes but do not cause symptoms. The functions that allow *Cmm* infection and colonization of its tomato host, resulting in subsequent symptom expression, are encoded by genes on the *Cmm* chromosome.

Recently, BLI was applied to assess colonization dynamics of this pathogen. It was thought that the merits of BLI would help to shed light on the dynamics of seed to seedling transmission of *Cmm* and its localization in the plant in real time. Here, we discuss the generation of bioluminescent *Cmm* and findings garnered from the use of BLI to monitor this pathogen.

4.3 Construction of Bioluminescent **C. michiganensis** *subsp.* **michiganensis** *Strains*

The generation of bioluminescent *Cmm* was facilitated by a highly efficient transposon mutagenesis system that was described by Kirchner et al. [40]. This system uses Tn*1409* and was modified to insert the *P. luminescens lux* operon into the genome of *Cmm* [79]. The construction and screening of bioluminescent *Cmm* is described in brief in the next section.

4.3.1 Construction of Plasmids Carrying the *lux* Operon

Three different vectors were constructed to insert the *lux* operon into *Cmm*. Specifically, a promoter-less lux operon (luxABCDE) in a vector pXen5 (Xenogen Corporation, USA) was modified by adding a Gram-positive bacterial ribosome binding site (AGGAGG) upstream of each *lux* gene for optimal expression in Gram-positive bacteria [24]. To construct pXX1, the modified *lux* operon and a kanamycin resistance gene from pXen5 were amplified, enzymatically digested, and ligated into a similarly digested Tn*5* transposon construction vector, pMod 3 (Epicentre, USA). The vector pXX1 carries an EZ::TN⟨*lux-kan*⟩ cassette, which is flanked by the Tn5 transposon mosaic ends.

Another vector was constructed by deploying inverse polymerase chain reaction (PCR) on the Tn*1409* transposon vector pKGT452Cβ [40] to amplify all but the region for the *lux* insertion. The inverse PCR product was digested and ligated to the similarly digested *lux* operon from pXen5, generating pXX2 (chloramphenicol resistance gene::*luxABCDE*::Tn*1409*). A similar vector, pXX3, was designed using a nonmodified (Gram-negative) promoterless *lux* operon and carries chloramphenicol resistance gene::*luxCDABE*::Tn*1409*. The three vectors (pXX1, pXX2, and pXX3) were cloned in *E. coli* DH5α and propagated in the *dam*- and *dcm*-deficient *E. coli* strain, ER2925, because the use of unmethylated DNA improves transformation efficiency in *Cmm* [40].

4.3.2 Mutagenesis and Isolation of Bioluminescent *C. michiganensis* subsp. *michiganensis*

Bioluminescent *Cmm* strains were successfully generated using pXX2, the suicide vector containing the modified transposon Tn*1409* (Fig. 6a). In brief, electrocompetent cells were prepared by growing *Cmm* to OD_{600} of 0.5–0.7 in TBY broth (Tryptone broth with yeast) at room temperature. The cells were then pretreated with 2.5 % glycine for 2 h, harvested, and washed three times with sterile ice-cold distilled water and twice with 10 % glycerol. The pellets were then resuspended in 15 % glycerol to 1/250 of the original volume. Transformation was achieved by mixing 100 μl of electrocompetent *Cmm* cells with 2 μl of the EZ::TN transposome complex or 1 μl (1 μg) of the plasmid pXX2, and electroporation (2.5 kV; 25 μF; 600 Ω) was performed in 0.2-cm cuvettes. The cells were then mixed with 0.4 ml of SB (Super Broth) medium and incubated with shaking (140 rpm) for 3 h at 28 °C. The transformed cells were spread onto SB agar plates containing chloramphenicol and incubated for 4–7 days at 28 °C. The resulting colonies were recovered and streaked onto NBY (Nutrient broth-yeast extract broth) agar plates containing antibiotics and screened for bioluminescence using the IVIS (Fig. 6b).

The insertion of Tn*1409* in the genome of *Cmm* is random [40], and the mutagenesis might affect virulence genes and the pathogenesis of the bioluminescent strain. Furthermore, the promoterless *lux* operon may be inserted downstream of promoters of different expression levels, which in turn would affect the strength of the bioluminescent signal. Subsequently, it is important to screen the bioluminescent *Cmm* colonies and choose the most suitable strain. One strain (BL-Cmm17) possessed strong and constitutive bioluminescent expression and growth and virulence properties that were similar to the wild-type strain (C290). Furthermore, BL-Cmm17 retained antibiotic resistance and bioluminescence, even after five

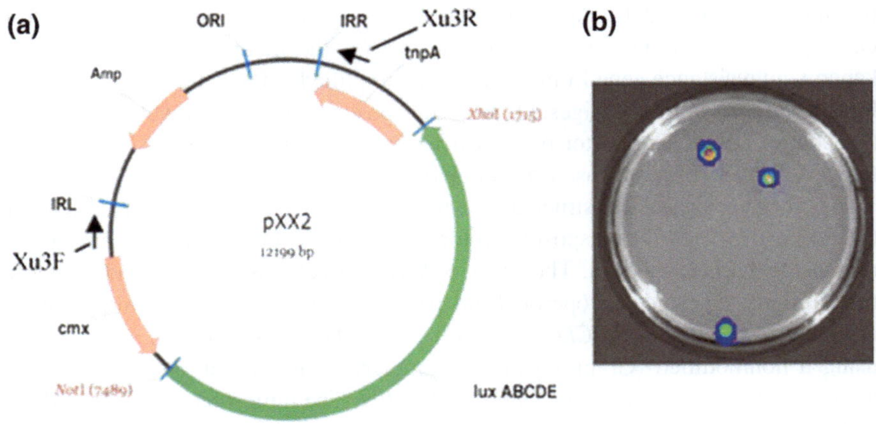

Fig. 6 **a** Map of the Tn*1409* transposon vector, pXX2, which was used to generate the bioluminescent *Clavibacter michiganensis* subsp. *michiganensis* strain, BL-Cmm17. *Amp* ampicillin resistance gene; *cmx* chloramphenicol resistance gene; *tnpA* transposase IS*1409*; *IRR* right inverted repeat; *IRL* left inverted repeat; and *ORI* origin of replication. The *arrows* indicate locations of the primers (Xu3F and 3 R) used for sequencing. **b** Bioluminescent colonies detected after the insertion of pXX2 into the genome of *Clavibacter*. Screening colonies for bioluminescence was performed using the IVIS machine [79]

rounds of nonselective growth for approximately 50 generations, indicating that the *lux* insertion was stable. Based on the stability of the integrated *lux* operon and its constitutive expression, as well as the wild-type-like growth and virulence properties of BL-Cmm17, it was concluded that the *lux* operon was inserted downstream of a strong promoter without apparently or negatively affecting any housekeeping or essential virulence genes. Hence, BL-Cmm17 was selected for BLI analysis of *Cmm* infections [79].

4.4 Knowledge Garnered from the Use of Bioluminescence to Study **C. michiganensis** *subsp.* **michiganensis** *in Tomato Plants and Seeds*

BLI was applied to monitor the colonization of *Cmm* in germinating tomato seeds. For this purpose, healthy tomato seeds were soaked in a solution containing BL-Cmm17 (10^8 CFU/ml) for 5 min under vacuum. The infected seeds were then air-dried and placed on water agar (1 %) in a square petri dish (10 by 10 cm, with a 13-mm grid) and incubated at 25 °C in the dark [79]. Daily monitoring over 5 days showed bioluminescence on the seed coats. Signals were also observed on some of the radicles that punctured the seed coats on day 3 postinoculation (DPI). As germination progressed, bioluminescence was detected on the hypocotyls and cotyledons. The latter locations and the seed coats harbored stronger signals than those on the radicles. These observations were confirmed by quantification of *Cmm* CFU in germinating seeds.

Specifically, up to 6 and 5 logs of bacteria were detected on the seed coats during the first 3 days postinoculation and on day 5, respectively, while carefully dissected and aseptically collected hypocotyls, cotyledons, and radicles contained an average of 4.5–5.4 logs of bacteria at day 5 [79]. Therefore, BLI revealed that BL-Cmm17 aggregated on the hypocotyls and cotyledons of the seedlings. This is an important observation considering that the epiphytic *Cmm* can easily be spread through irrigation and leaf-to-leaf contact, a significant route of transmission in greenhouse transplant production where the density of seedlings is very high [80]. BLI revealed that the localization in germinating seeds was possibly optimal for the spread of *Cmm* and the onset of future infections. These conclusions are supported by other studies that reported early canker symptoms on the cotyledons and stems, while *Cmm* populations were associated with tomato leaves [4, 29, 68].

Xu et al. [80] used BLI to monitor *Cmm* colonization of tomato seedlings 5, 10, 15, and 20 days postinoculation under different relative humidity (RH) treatments (83 ± 5 % designated high RH vs. 45 ± 12 % as low RH). Roots were similarly analyzed, imaged every 5 days postinoculation after sacrificing one randomly selected seedling per replicate per treatment. BLI showed that strong bioluminescence appeared in the cotyledon petioles at 5 DPI (Fig. 7). Signals observed in stems indicated the initiation of stem colonization, which was obvious at day 10 DPI in seedlings maintained in both low and high RH. At the same time, light was detected in the petioles of the first and second true leaves in tomato seedlings under high RH. The bioluminescent signals quickly spread to both roots and shoots.

At 15 DPI, bioluminescence was observed in lower leaves of tomato seedlings under low RH and in all leaves of seedlings under high RH. However, signals were detected throughout the entire seedling at 20 DPI regardless of the RH treatment. This corresponded with slight wilting of the first true leaf of seedlings maintained under high RH, while wilting was more obvious in leaves of seedlings under both low and high RH by 20 DPI. Notably, higher bioluminescence was observed at the nodes in stems as compared to internodes, while in individual leaves, the strongest signals appeared in the petiole nearest the main stem at 20 DPI.

These results highlight the invasive nature of *Cmm*. Combining BLI with identification of symptoms showed the rapid onset of infections in seedlings. At 20 DPI, BLI showed that the seedling was almost completely colonized down to the roots. Therefore, rapid colonization of seedlings indicates that control of this disease must focus at the level of the seeds; however, it would be interesting to use BLI and bioluminescent *Cmm* to assess possible bactericides during the course of infection. Taken together, these studies highlight the usefulness of BLI to monitor *Cmm* in real time during the life cycle of the infected plant and under different environmental conditions.

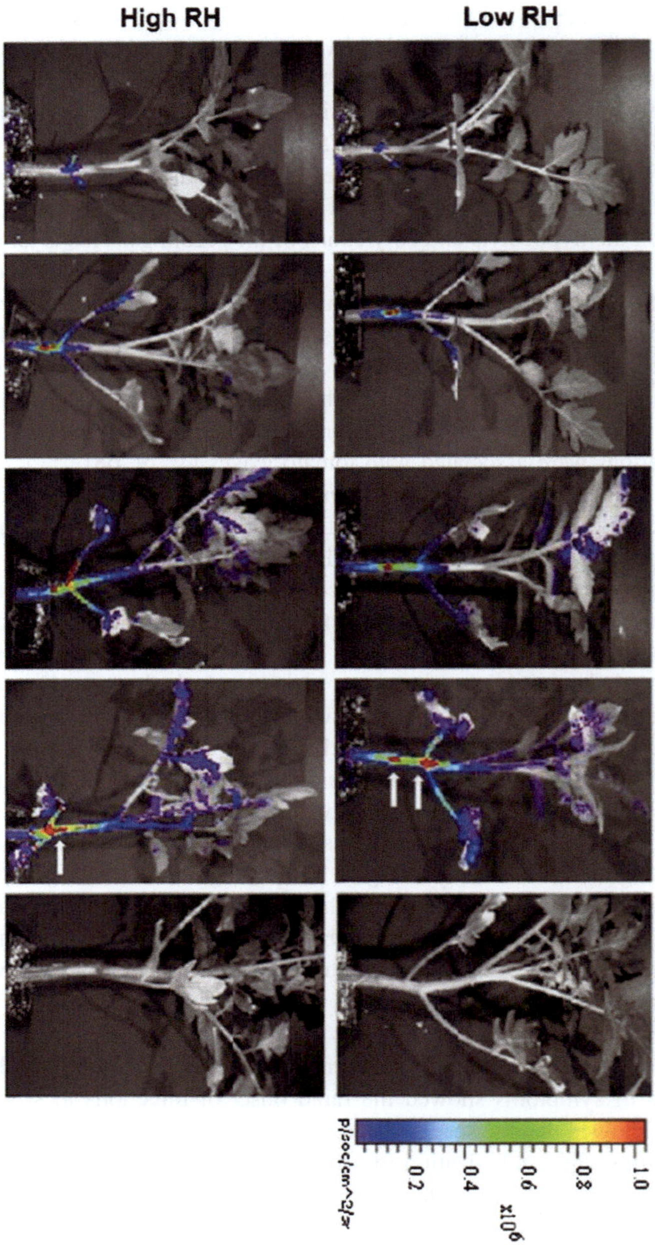

Fig. 7 BLI of the localization of the bioluminescent *Clavibacter michiganensis* subsp. *michiganensis* strain BL-Cmm17 in tomato seedlings under low and high relative humidity (RH). The days postinoculation (DPI) are represented on the top of each panel. Control seedlings (CK) that were inoculated with only water are also imaged at 20 DPI using the IVIS machine. The arrows point to strong bioluminescent signals at the nodes. The rainbow scale represents light intensity [80]

5 The Promise of Light, *Video Vidi Visum*: Conclusions and Prospect

A first landmark for the application of BLI in bacterial pathogenesis was the use of a non-attenuated bioluminescent *Salmonella typhimurium* to monitor disease progression in mice [10]. Since then, BLI has become a significant tool for studying a plethora of significant bacterial pathogens in live animal models. This successful, real-time, and noninvasive application in many instances has provided superior insights to temporal and spatial localization of pathogens, pathogenesis, infections, host responses, and therapeutic interventions as compared to the conventional time-of-death (end-point) analysis. It is perhaps the greatest virtue of BLI that the aforementioned advances occur while minimizing the numbers and suffering of animals, reducing and focusing labor, and enhancing our understanding of infectious diseases. With the advancement of detection technology and molecular methodology, in general, we predict that BLI will be more widely adopted by researchers in infectious diseases.

However, there are still many hurdles that BLI technology needs to surpass in order to become more useful. For example, the use of large animal models is limited by the size of the imaging system as well as the sensitivity of light detection of the CCD camera. The latter is relevant because the depth of the infected tissue in the host is an obvious limiting factor. There is also the issue of the compatibility of the available or popular *lux* systems with different bacterial pathogens [3]. It is already known that the bioluminescence associated with the *lux* operon requires oxygen, which impedes the application of BLI to study anaerobic pathogens or survival in oxygen-limited environments. Another question is also the ability to generate bioluminescent strains of metabolically fastidious and genetically difficult to manipulate bacterial pathogens. For example, our experimentation with the use of the *lux* system from *P. luminescens* in *Campylobacter jejuni*, a fastidious and microaerobic food-borne bacterial pathogen, resulted in bioluminescent strains [38]. However, the cognate bioluminescent signals were not consistently detected in chicks, the animal model for *C. jejuni*, or piglets even after enrichment or aeration of harvested organs (Kassem and Rajahsekara, unpublished data). This may have been due to multiple factors, including pathogen physiology, unsuitability of the *lux* system, and/or the mutagenesis approach used to generate the bioluminescent strains.

The shortcomings and the successes described here can also be generally applied to BLI of bacterial pathogens in the plant host. However, BLI in these hosts can be severely compounded by natural plant phosphorescence, which could overwhelm light signals from the bioluminescent bacteria [80]. This can be addressed by a simple approach: incubation of plants in the dark for 10 min quenches the background phosphorescence [80]. Alternatively, special modules can be inserted into the IVIS to eliminate the background.

In conclusion, BLI is a promising tool that can yield novel and significant insights into bacterial pathogen–host interactions. Future advancements will enhance its sensitivity and applicability to meet the demands of an important field.

Acknowledgment Research in the Rajashekara and Miller laboratories is supported by funds from the Agriculture and Food Research Initiative (AFRI, USDA), the Specialty Crops Research Initiative (SCRI-2010-600-25320, USDA), the Ohio Agricultural Research and Development Center, and the Ohio State University.

References

1. Alexander B, Schnurrenberger PR, Brown RR (1981) Numbers of *Brucella abortus* in the placenta, umbilicus and fetal fluid of two naturally infected cows. Vet Rec 108:500
2. Allen MS, Wilgus JR, Chewning CS, Sayler GS, Simpson ML (2007) A destabilized bacterial luciferase for dynamic gene expression studies. Syst Synth Biol 1:3–9
3. Andreu N, Zelmer A, Wiles S (2011) Noninvasive biophotonic imaging for studies of infectious disease. FEMS Microbiol Rev 35:360–394
4. Carlton WM, Gleason ML, Braun EJ (1998) Ingress of *Clavibacter michiganensis* subsp. *michiganensis* into tomato leaves through hydathodes. Phytopathology 88:525–529
5. Chabot R, Kloepper JW, Antoun H, Beauchamp CJ (1996) Root colonization of maize and lettuce by bioluminescent *Rhizobium leguminosarum* biovar *phaseoli*. Appl Environ Microbiol 62:2767–2772
6. Chalupowicz L, Zellermann E-M, Fluegel M, Dror O, Eichenlaub R, Gartemann K-H, Savidor A, Sessa G, Iraki N, Barash I, Manulis-Sasson S (2012) Colonization and movement of GFP labeled *Clavibactermichiganensis* subsp. *michiganensis* during tomato infection. Phytopathology 102:23–31
7. Chang RJ, Ries SM, Pataky JK (1991) Dissemination of *Clavibacter michiganensis* subsp. *michiganensis* by practices used to produce tomato transplants. Phytopathology 81:1276–1281
8. Cirvilleri G, Lindow SE (1994) Differential expression of genes of *Pseudomonas syringae* on leaves and in culture evaluated with random genomic *lux* fusions. Mol Ecol 3:249–257
9. Cirvilleri G, BellaP CataraV (2000) Luciferase genes as a marker for *Pseudomonas corrugata*. J Plant Pathol 82:237–241
10. Contag CH, Contag PR, Mullins JI, Spilman SD, Stevenson DK, Benaron DA (1995) Photonic detection of bacterial pathogens in living hosts. Mol Microbiol 18:593–603
11. Contag PR (2008) Bioluminescence imaging to evaluate infections and host response in vivo. Methods Mol Biol 415:101–118
12. Corbel M (1997) Brucellosis: an overview. Emerg Infect Dis 3:213–221
13. Corish P, Tyler-Smith C (1999) Attenuation of green fluorescent protein half-life in mammalian cells. Protein Eng 12:1035–1040
14. Delrue RMM-LM, Lestrate P, Danese I, Bielarz V, Mertens P, De Bolle X, Tibor A, Gorvel JP, Letesson JJ (2001) Identification of *Brucella* spp. genes involved in intracellular trafficking. Cell Microbiol 3:487–497
15. De Weger LA, DunbarP MahafeeWF, LugtenbergJJ SaylerGS (1991) Use of bioluminescent markers to detect *Pseudomonas* spp. in the rhizosphere. Appl Environ Microbiol 57:3641–3644
16. Engebrecht J, Silverman M (1984) Identification of genes and gene products necessary for bacterial bioluminescence. Proc Natl Acad Sci USA 81:4154–4158
17. Eichenlaub R, Gartemann KH (2011) The *Clavibacter michiganensis* subspecies, molecular investigation of Gram-positive bacterial plant pathogens. Annu Rev Phytopathol 49:445–464
18. Essenberg R, Seshadri R, Nelson K, Paulsen I (2002) Sugar metabolism by *Brucellae*. Vet Microbiol 90:249–261
19. European and Mediterranean Plant Protection Organization (2005) *Clavibacter michiganensis* subsp. *michiganensis*. EPPO Bull 35:275–283

20. Fan J, Lamb C, Crooks C (2008) High-throughput quantitative luminescence assay of the growth in planta of *Pseudomonas syringae* chromosomally tagged with *Photorhabdus luminescens luxCDABE*. The Plant J 53:393–399
21. Fatmi M, Bolkan HA, Schaad NW (1991) Seed treatments for eradicating *Clavibacter michiganensis* subsp. *michiganensis* from naturally infected tomato seeds. Plant Dis 75:383–385
22. Foulongne VBG, Cazevieille C, Michaux-Charachon S, O'Callaghan D (2000) Identification of *Brucella suis* genes affecting intracellular survival in an in vitro human macrophage infection model by signature-tagged transposon mutagenesis. Infect Immun 68:1297–1303
23. Frackman S, Anhalt M, Nealson KH (1990) Cloning, organization, and expression of the bioluminescence genes of *Xenorhabdus luminescens*. J Bacteriol 172:5767–5773
24. Francis KP, Bellinger-Kawahara C, Joe D, Purchio TF, Hawkinson MJ, Contag PR (2000) Monitoring bioluminescent *Staphylococcus aureus* infections in living mice using a novel *luxABCDE*construction. Infect Immun 68:3594–3600
25. Franz DR (1999) Foreign animal disease agents as weapons in biological warfare. Ann NY Acad Sci 894:100–104
26. Fukui R, McElhaney R, Fukui H, Alvarez AM, Nelson SC (1996) Relationship between symptom development and actual sites of infection in leaves of anthurium inoculated with a bioluminescent strain of *Xanthomonas campestris* pv. *dieffenbachiae*. Appl Environ Microbiol 62:1021–1028
27. Gartemann KH, Kirchner O, Engemann J, Gräfen I, Eichenlaub R, Burger A (2003) *Clavibacter michiganensis* subsp. *michiganensis*: first steps in the understanding of virulence of a Gram-positive phytopathogenic bacterium. J Biotechnol 106:179–191
28. Gitaitis RD, Beaver RW, Voloudakis AE (1991) Detection of *Clavibacter michiganensis* subsp. *michiganensis* in symptomless tomato transplants. Plant Dis 75:834–838
29. Gleason ML, Ricker MD, Gitaitis RD (1993) Recent progress in understanding and controlling bacterial canker of tomato in Eastern North America. Plant Dis 77:1069–1076
30. Glomski IJ, Piris-Gimenez A, Huerre M, Mock M, Goossens PL (2007) Primaryinvolvement of pharynx and peyer's patch in inhalational and intestinal anthrax. PLoS Pathog 3:e76
31. Glynn M, Lynn TV (2008) Brucellosis. J Amer Vet Med Assoc 233:900–908
32. Greer LF III, Szalay AA (2002) Imaging of light emission from the expression ofluciferases in living cells and organisms: a review. Luminescence 17:43–74
33. Griffiths MW (2000) How novel methods can help discover more information about foodborne pathogens. Can J Infect Dis 11:142–153
34. Hardy J, Francis KP, DeBoer M, Chu P, Gibbs K, Contag CH (2004) Extracellularreplication of *Listeria monocytogenes* in the murine gall bladder. Science 303:851–853
35. Hong PCTR, Ficht TA (2000) Identification of genes required for chronic persistence of *Brucella abortus* in mice. Infect Immun 68:4102–4107
36. Jahr H, Burger A, Bahro R, Eichenlaub R, Ahlemeyer J (1999) Interactions between *Clavibacter michiganensis* and its host plants. Environ Microbiol 1:113–118
37. Kamoun S, Kado CI (1990) A plant-inducible gene of *Xanthomonas campestris* pv. *campestris* encodes an exocellular component required for growth in the host and hypersensitivity on nonhosts. J Bacteriol 172:5165–5172
38. Kassem II, Sanad Y, Gangaiah D, Lilburn M, Lejeune J, Rajashekara G (2010) Use of bioluminescence imaging to monitor *Campylobacter* survival in chicken litter. J Appl Microbiol 109:1988–1997
39. Kaufmann AF, Meltzer MI, Schmid GP (1997) The economic impact of a bioterrorist attack: are prevention and postattack intervention programs justifiable? Emerg Infect Dis 3:83–94
40. Kirchner O, Gartemann KH, Zellermann EM, Eichenlaub R, Burger A (2001) A highlyefficient transposon mutagenesis system for the tomato pathogen *Clavibacter michiganensis* subsp. *michiganensis*. Mol Plant Microbe Interact 14:1312–1318
41. Ko J, Gendron-Fitzpatrick A, Ficht TA, Splitter GA (2002) Virulence criteria for *Brucella abortus* strains as determined by interferon regulatory factor 1-deficient mice. Infect Immun 70:7004–7012

42. Ko J, Gendron-Fitzpatrick A, Splitter GA (2002) Susceptibility of IFN regulatory factor-1 and IFN consensus sequence binding protein-deficient mice to brucellosis. J Immunol 168:2433–2440
43. Koc Z, Turunc T, Boga C (2007) Gonadal brucellar abscess: imaging and clinical findings in 3 cases and review of the literature. J Clin Ultrasound 35:395–400
44. Kulkarni R, Chunchanur SK, Ajantha GS, Shubhada C, Jain P (2009) Presumptive diagnosis of *brucella* epididymoorchitis by modified cold ZN staining of testicular pus sample. Indian J Med Res 130:484–486
45. Li Z, Szittner R, Meighen EA (1993) Subunit interactions and the role of the *luxA* polypeptide in controlling thermal stability and catalytic properties in recombinant luciferase hybrids. Biochim Biophys Acta 1158:137–145
46. Maudlin I, Weber S (2006) The control of neglected zoonotic diseases: a route to poverty alleviation. WHO, Geneva
47. Meighen E (1993) Bacterial bioluminescence: organization, regulation, and application of the *lux* genes. FASEB J 7:1016–1022
48. Meletzus D, Dreier J, Bermpohl A, Eichenlaub R (1993) Evidence for plasmid-encoded virulence factors in the phytopathogenic bacterium *Clavibacter michiganensis* subsp. *michiganensis* NCPPB382. J Bacteriol 175:2131–2136
49. Miller SA, Rowe RC, Riedel RM (1996) Bacterial spot, speck, and canker of tomatoes HYG-3120-96. Ohio State Univ Extension
50. O'Callaghan DCC, Allardet-Servent A, Boschiroli ML, Bourg G, Foulongne V, Frutos P, Kulakov Y, Ramuz M (1999) A homologue of the *Agrobacterium tumefaciens VirB* and B*ordetella pertussis* Ptl type IV secretion systems is essential for intracellular survival of *Brucella suis*. Mol Microbiol 33:1210–1220
51. Paynter CD, Arnold DL, Salisbury VC, Jackson RW (2006) The use of bioluminescence for monitoring in planta growth dynamics of a *Pseudomonas syringae* plant pathogen. Eur J Plant Pathol 115:363–366
52. Petersen E, Rajashekara G, Sanakkayala N, Eskra L, Harms J, Splitter G (2013) Erythritol triggers expression of virulence traits in *Brucella melitensis*. Microbes Infect 15:440–449
53. Radhakrishnan GKYQ, Harms JS, Splitter GA (2009) *Brucella* TIR domain-containing protein mimics properties of the toll-like receptor adaptor protein TIRAP. J Biol Chem 284:9892–9898
54. Rajashekara G, Glover DA, Banai M, O'Callaghan D, Splitter GA (2006) Attenuated bioluminescent *Brucella melitensis* mutants GR019 (*virB4*), GR024 (*galE*), and GR026 (*BMEI1090-BMEI1091*) confer protection in mice. Infect Immun 74:2925–2936
55. Rajashekara G, Glover DA, Krepps M, Splitter GA (2005) Temporal analysis of pathogenic events in virulent and avirulent *Brucella melitensis* infections. Cell Microbiol 7:1459–1473
56. Rajashekara G, Krepps M, Eskra L, Mathison A, Montgomery A, Ishii Y, Splitter G (2005) Unraveling *Brucella* genomics and pathogenesis in immunocompromised IRF-1-/- mice. Am J Reprod Immunol 54:358–368
57. Ricker MD, Riedel RM (1993) Effect of secondary spread of *Clavibacter michiganensis* subsp. *michiganensis* on yield of northern processing tomatoes. Plant Dis 77:364–366
58. Ritchie JM, Campbell GR, Shepherd J, Beaton Y, Jones D, Killham K, Artz RR (2003) A stable bioluminescent construct of *Escherichia coli* O157:H7 for hazard assessments of long-term survival in the environment. Appl Environ Microbiol 69:3359–3367
59. Samartino L, Enright FM (1993) Pathogenesis of abortion of bovine brucellosis. Comp Immunol Microbiol Infect Dis 16:95–101
60. Shaner NC, Steinbach PA, Tsien RY (2005) A guide to choosing fluorescent proteins. Nat Methods 2:905–909
61. Sharabani G, Manulis-Sasson S, Borenstein M, Shulhani R, Chalupowicz L, Lofthouse M, Shtienberg D (2013) The significance of guttation inthe secondary spread of *Clavibactermichiganensis* in tomato greenhouses. Plant Pathol 62:578–586
62. Sharabani G, Shtienberg D, Borenstein M, Shulhani R, Sofer M, Lofthouse M, Chalupowicz L, Barel V, Manulis-Sasson S (2012) Effects of plant age on disease development and virulence of Clavibacter michiganensis subsp. michiganensis on tomato. Plant Pathology 62:1114–1122

63. Shaw JJ, DaneF GeigerD, Kloepper JW (1992) Use of bioluminescence for detection of genetically engineered microorganisms released into the environment. Appl Environ Microbiol 58:267–273
64. Shaw JJ, Kado CI, Settles LG (1988) Transposon Tn*4431* mutagenesis of *Xanthomonas campestris* pv. *campestris*: characterization of a non pathogenic mutant and cloning of a locus for pathogenicity. Mol Plant-Microbe Interact 1:39–45
65. Shirakawa T, Sasaki T, Ozaki K (1991) Ecology and control of tomato bacterial canker and detection methods of its pathogen. Jpn Agric Res Q 25:27–32
66. Sieira R, Comerci DJ, Sanchez DO, Ugalde RA (2000) A homologue of an operon required for DNA transfer in *Agrobacterium* is required in *Brucella abortus* for virulence and intracellular multiplication. J Bacteriol 182:4849–4855
67. Smith H, Williams AE, Pearce JH, Keppie J, Harris-Smith PW, Fitz-George RB, Witt K (1962) Foetal erythritol: a cause of the localization of *Brucella abortus* in bovine contagious abortion. Nature 193:47–49
68. Strider DL (1969) Bacterial canker of tomato, a literature review and bibliography. NC Agric Exp Stn Tech Bull 193:1–80
69. Szittner R, Meighen E (1990) Nucleotide sequence, expression, and properties of luciferase coded by *lux* genes from a terrestrial bacterium. J Biol Chem 265:16581–16587
70. Tancos MA, Chalupowicz L, Barash I, Manulis-Sasson S, Smart CD (2013) Tomato fruit and seed colonization by *Clavibacter michiganensis* subsp. *michiganensis* through external and internal routes. Appl Env Microbiol 79:6948–6957
71. Thorn RM, Nelson SM, Greenman J (2007) Use of a bioluminescent *Pseudomonas aeruginosa* strain within an in vitro microbiological system, as a model of wound infection, to assess the antimicrobial efficacy of wound dressings by monitoring light production. Antimicrob Agents Chemother 51:3217–3224
72. Tsiantos J (1987) Transmission of bacterium *Corynebacterium michiganense* pv. *michiganense* by seeds. J Phytopathology 119:142–146
73. Tsien RY (1998) The green fluorescent protein. Annu Rev Biochem 67:509–544
74. Vallejo JG, Stevens AM, Dutton RV, Kaplan SL (1996) Hepatosplenic abscesses due to *Brucella melitensis*: report of a case involving a child and review of the literature. Clin Infect Dis 22:485–489
75. Verkhusha VV, Lukyanov KA (2004) The molecular properties and applications of Anthozoa fluorescent proteins and chromoproteins. Nat Biotechnol 22:289–296
76. Waidmann MS, Bleichrodt FS, Laslo T, Riedel CU (2011) Bacterial luciferasereporters: the Swiss army knife of molecular biology. Bioeng Bugs 2:8–16
77. Widder EA (2010) Bioluminescence in the ocean: origins of biological, chemical, and ecological diversity. Science 328:704–708
78. Wiles S, Clare S, Harker J, Huett A, Young D, Dougan G, Frankel G (2004) Organ specificity, colonization and clearance dynamics in vivo following oral challenges with the murine pathogen Citrobacter rodentium. Cell Microbiol 6:963–972
79. Xu X, Miller SA, Baysal-Gurel F, Gartemann KH, Eichenlaub R, Rajashekara G (2010) Bioluminescence imaging of *Clavibacter michiganensis* subsp. *michiganensis* infection of tomato seeds and plants. Appl Environ Microbiol 76:3978–3988
80. Xu X, Rajashekara G, Paul PA, Miller SA (2012) Colonization of tomato seedlings by bioluminescent *Clavibacter michiganensis* subsp. *michiganensis* under different humidity regimes. Phytopathology. 102:177–184
81. Young E (1995) An overview of human brucellosis. Clin Infect Dis 21:283–289

Part IV
Applications of Bioluminescence in Health

Adv Biochem Eng Biotechnol (2016) 154: 149–180
DOI 10.1007/10_2015_336

Published Online: 17 October 2015

Enabling Aequorin for Biotechnology Applications Through Genetic Engineering

Kristen Grinstead, Smita Joel, Jean-Marc Zingg, Emre Dikici and Sylvia Daunert

Abstract In recent years, luminescent proteins have been studied for their potential application in a variety of detection systems. Bioluminescent proteins, which do not require an external excitation source, are especially well-suited as reporters in analytical detection. The photoprotein aequorin is a bioluminescent protein that can be engineered for use as a molecular reporter under a wide range of conditions while maintaining its sensitivity. Herein, the characteristics of aequorin as well as the engineering and production of aequorin variants and their impact on signal detection in biological systems are presented. The structural features and activity of aequorin, its benefits as a label for sensing and applications in highly sensitive detection, as well as in gaining insight into biological processes are discussed. Among those, focus has been placed on the highly sensitive calcium detection in vivo, in vitro DNA and small molecule sensing, and development of in vivo imaging technologies.

K. Grinstead · S. Joel · J.-M. Zingg · E. Dikici · S. Daunert (✉)
Department of Biochemistry and Molecular Biology, Miller School of Medicine, University of Miami, Miami, FL 33136, USA
e-mail: SDaunert@med.miami.edu

Graphical Abstract

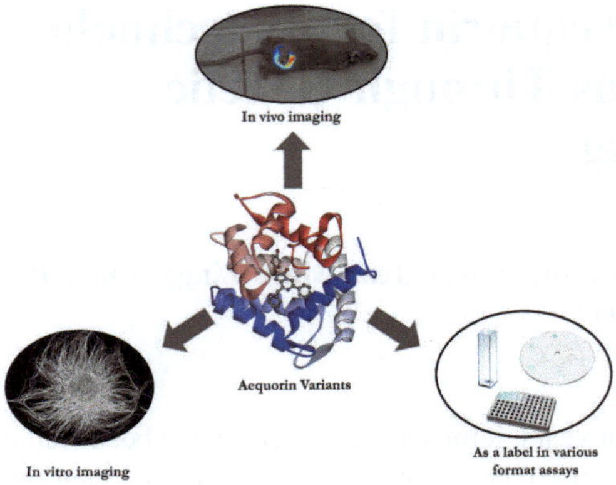

Keywords Aequorin · Bioluminescence · Calcium · ELISA · Imaging · Mutagenesis · Whole-cell assays

Contents

1 Introduction

During the last three decades, the field of biotechnology experienced an increase in the demand for the ultrasensitive detection of biomolecules both in ex vivo as well as in vivo applications, pointing out a need for novel molecular reporters capable of providing high sensitivity and low detection limits for analytical systems [1, 2].

The search for signaling molecules to meet this demand led to the discovery of luminescent proteins such as green fluorescent protein (GFP). The identification, isolation, and cloning of GFP was an important milestone in biotechnology as recognized by the award of the 2008 Nobel Prize in Chemistry to Shimomura, Tsien, and Chalfie for discovering and enabling the use of GFP in biological applications [3, 4]. The use of GFP as a molecular reporter allowed scientists to image and monitor in vivo cellular and molecular processes for the first time [5]. Although fluorescent proteins such as GFP have been used as molecular reporters in several analytical applications, as with any other fluorescent reporter, GFP needs to be excited by an external source to emit light. The external excitation causes emission of light not only from GFP, but also from other fluorescent molecules in the environment, thus causing the production of a background signal that compromises the sensitivity of the detection afforded by GFP. Other factors that limit the sensitivity of fluorescence-based detection systems include photobleaching, phototoxicity, and autofluorescence [5, 6].

Bioluminescence, the phenomenon wherein light is emitted by living organisms (Fig. 1) [7] and its use in biotechnology as a detection system has emerged as a powerful alternative to fluorescence. Bioluminescence occurs in a wide variety of organisms, from simple one-cell bacteria and sea squirts to octopodes and fireflies, where it is employed for an equally wide variety of purposes, from luring prey and attracting mates, to use as a decoy and stealth shield [6, 8–14]. The emission of light in all bioluminescent organisms is of a biochemical nature and involves a reaction that is catalyzed either by an enzyme or a bioluminescent protein. Examples of such proteins found in nature include bacterial or firefly luciferase, clytin, obelin, and aequorin [2, 4, 15–18]. Aequorin, a photoprotein found in bioluminescent jellyfish, is active at physiological pH, nontoxic, environmentally harmless, and has been cloned for expression in bacteria, plants, and animals. It has been widely used in calcium detection for decades, and is the reporter of choice in high-throughput assays for GPCRs, ion channels, and tyrosine kinase receptors [19–25]. Given the wealth of the body of work in the last 20 years with bioluminescence, herein we limit our discussion to the most recent impact of aequorin on signal detection, including monitoring calcium fluctuations in living cells, the detection of small molecules via aequorin chimeras, and in vivo imaging.

Aequorin consists of a calcium-binding apoprotein that endows the protein with its structural characteristics, and an external organic luciferin, coelenterazine. The photoprotein also has four highly conserved EF-hands where calcium ions bind. The emission of light requires that coelenterazine become complexed with the apoaequorin to form the holoprotein and the binding of calcium ions. This binding triggers a change in conformation that results in an internal reaction ultimately leading to the emission of light. Thus, both calcium and coelenterazine are needed for the emission of bioluminescent signal.

The aequorin apoprotein is 189 amino acids long and has four EF-hands consisting of a helix–turn–helix motif. Three of the EF-hands are capable of binding calcium ions and often bind cooperatively; that is, the first binding of one of the EF-hands to calcium facilitates the binding of the next. The four EF-hands surround

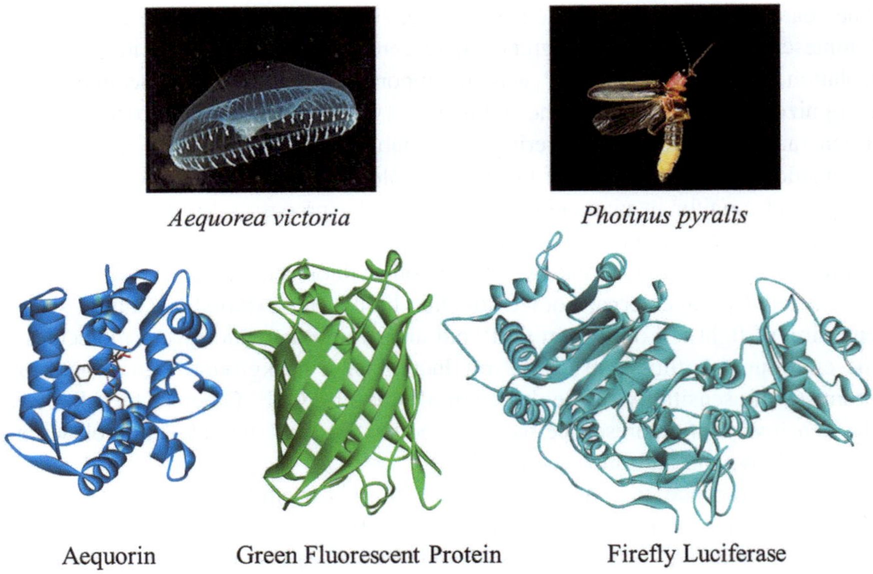

Fig. 1 Bioluminescent organisms and their proteins. The wavelength of emission of the protein determines the colors of the bioluminescent emission observed in the organisms. Crystal Jelly (*Aequorea victoria*): Copyright Sierra Blakely. Big Dipper Firefly (*Photinus pyralis*): Copyright Terry Priest. (Aequorin PDB Code: 1EJ3, GFP PDB Code: 1EMA, Luciferase PDB Code: 1LCI using DSViewer 3.5)

a hydrophobic pocket that contains three Tyr–His–Trp stabilizing triads that bind to the incorporated coelenterazine chromophore via a series of H-bonds [26–28]. When incorporated into the hydrophobic pocket of the apoaequorin and in the presence of an oxygen molecule, coelenterazine forms a hydroperoxide that is stabilized by the Tyr184 on the "tail" on the C-terminus, an amino acid located in one of three key Tyr–His–Trp triads within the pocket, producing an excited coelenterazine-2-hydroperoxide [29, 30]. Upon binding of calcium, the interhelical angles change, disrupting the hydrogen-binding network and the stabilization of the peroxide (Fig. 2). This causes the complex to destabilize, thus driving a biochemical reaction that results in the generation of coelenteramide and CO_2 as reaction products and emission of a blue flash of light at 469 nm [7, 31]. The replacement of coelenteramide with fresh coelenterazine by diffusion is limited and lengthy in time and, therefore, it should be noted that the signal produced is dependent on the amount of aequorin, not the amount of coelenterazine.

The EF-hand binding motif endows aequorin with a high affinity for calcium ion binding; indeed, the K_d = 0.2–5 μM. The high affinity of aequorin to calcium and generation of a bioluminescent signal in response to calcium binding formed the basis for the development of biosensing systems for calcium detection [32, 33].

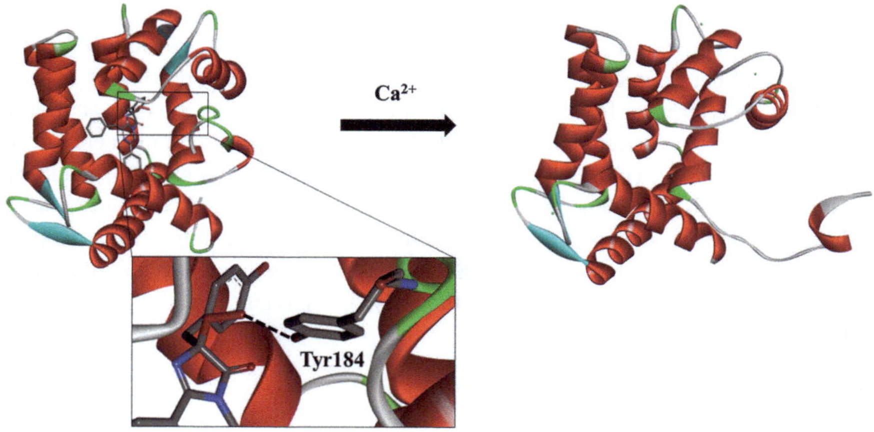

Fig. 2 Coelenterazine is shown in the hydrophobic pocket of aequorin. The hydrogen bond between the covalently bound oxygen molecule and neighboring tyrosine at position 184 is shown with a *dashed black line*. Oxygen atoms are in red, nitrogen atoms are in *dark blue*, and calcium ions are in *green* (Aequorin PDB Code: 1EJ3; Aequorin and Ca^{2+} PDB Code: 1SL8)

2 Aequorin in Calcium Detection

Calcium ions are abundant in cells and serve a critical role as activators of second messengers in development, metabolism, proliferation, and cell death. Although the calcium signaling pathways are complex and tailored to a cell's requirements, the function can be divided into two parts, the ON mechanism when calcium is released from internal stores into the cytoplasm, and the OFF mechanism, when calcium is removed from the cytoplasm and the resting state is restored. Disruption of calcium signaling is linked to a variety of severe health conditions, including neurodegeneration and heart disease [34–37]. Furthermore, the G-coupled protein receptor (GPCR), a major pharmaceutical target accounting for roughly 50 % of all pharmaceuticals on the market, releases calcium stores from the endoplasmic reticulum upon activation, setting off a chain of events within the cell, inhibiting or accelerating cellular processes through predetermined signaling pathways [38]. GPCRs can be found all throughout the body and are therapeutic targets in areas such as obesity, inflammation, pain, cardiac dysfunction, and cancer [39].

The intrinsic ability of aequorin to bind tightly to calcium led to the development of calcium detection systems within a few years of the photoprotein's initial discovery and characterization [40, 41]. Aequorin's ability to function as a highly sensitive calcium biosensor resulted in enabling technologies such as high-throughput drug screening for pharmaceuticals that target receptors and junctions that permit the exchange of calcium ions, such as GCPRs [42, 43] and connexin 43 (Cx43, GJA1) gap junctions, an emerging target for developmental disorders, and cardio and neuropathy [44]. The genetic manipulation of aequorin has yielded highly selective and sensitive in vivo calcium tracking systems, such as those facilitated by aequorin fusion proteins, including aequorin fusions that can be consistently localized in specific cellular regions. This has allowed for the precise detection and determination of calcium changes in these targeted areas of interest, free from interference from the rest of the cell. Moreover, several aequorin–fluorescent protein fusions have been produced where the bioluminescence resonance energy transfer (BRET) between the aequorin and the fluorescent protein can be utilized to visualize and/or measure simultaneous calcium concentrations in different organelles, which allows for detailed information on intracellular calcium flux under certain conditions [45, 46]. Such measurements have provided insight on the functions of both healthy cells and those used as models of pathogenic systems wherever calcium plays a role.

2.1 Engineering Aequorin for Wide Applications in Calcium Detection

The importance of calcium as a signaling molecule during cell function can be seen in the sheer number of studies on that subject, far too many to list herein, and is a frequent topic of reviews [34, 47–51]. Optimization of the use of aequorin in in vivo calcium determination led to the investigation of the effect of calcium ions on the emission of aequorin's bioluminescence. Specifically, studies were undertaken to elucidate and understand the mechanism by which aequorin binds calcium through its four EF-hand motifs.

The amino acids may be coordinated with the calcium ion via the side-chain, backbone carbonyl (Y^-), or through hydrogen binding with a water molecule (X^-).

Calcium binding proteins with the EF-hand motif are highly conserved in structure and found in diverse organisms, from prokaryotic to eukaryotic forms of life, including humans [52, 53]. The binding of this motif to calcium ions is controlled by electrostatic and thermodynamic forces that favor binding to calcium over other common small ions, such as potassium, cadmium, or magnesium, commonly present in biological samples [53–56].

The "turn" in the helix–turn–helix structure of the EF-hand is a short alpha-helix section where the calcium ion is bound in a bipyramidal conformation by six amino acids contained in a highly conserved twelve amino acid long peptide sequence (Table 1) [57]. The six amino acids interact with the calcium ion via H-bonds and are labeled according to their position along a three-dimensional *X*-, *Y*-, or *Z*-axis (Fig. 3), + (upstream) or − (downstream), of a glycine at position 6 that is in the center of the six amino acid sequence.

The binding of the calcium ion to one or more EF-hands causes a change in the angles of the helices that is determined by the interaction with nearby hands and amino acid sequence unique to each member of the family of calcium binding proteins [53]. It was found that of the four EF-hands in aequorin, three are capable of calcium binding EF-I, EF-III, and EF-IV, each with different affinities. Typically,

Table 1 Conserved amino acid sequences in the calcium binding EF-hands

EF-Loop Number	1	2	3	4	5	6	7	8	9	10	11	12
Coordinate	X^+		Y^+		Z^+		Y^-		X^-			Z^-
Most Common	D 100 %	K 29 %	D 76 %	G 56 %	N 52 %	G 98 %	T 23 %	I 68 %	D 32 %	F 23 %	E 29 %	E 92 %
EF-I 24-35	D	V	N	H	N	G	K	I	S	L	D	E
EF-III 117-128	D	K	D	Q	N	G	A	I	T	D	D	E
EF-VI 153-164	D	I	D	E	S	G	Q	L	D	V	D	E

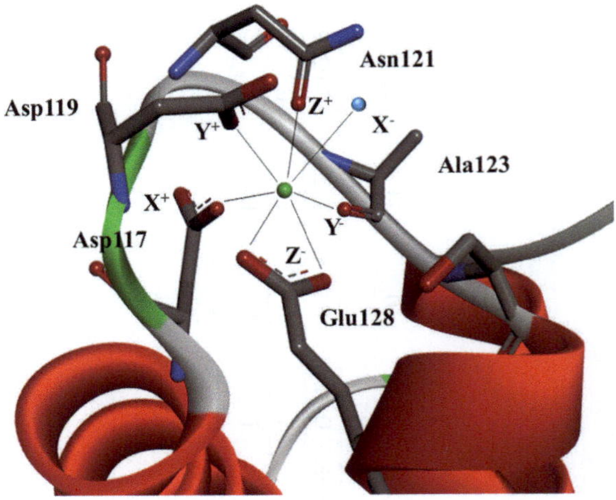

Fig. 3 Calcium coordinated within the EF-III hand of wild-type aequorin. Oxygen atoms are in red, nitrogen atoms are in *deep blue*, the calcium ion is in *green*, and the water molecule is in *light blue*. The coordinating bonds are shown by the *black lines* (calcium in aequorin PDB Code: 1SL8)

the bioluminescence is triggered by the binding of two calcium ions, one each to EF-I and EF-III. The fourth hand, EF-II, is not capable of calcium binding.

Studies have been conducted to determine the exact affinity of each individual binding EF-hand to calcium and how they interact with each other through the H-bond network [58]. The calcium binding capability of the EF-hands can be disrupted completely by removing a large nonpolar residue between two polar nonbinding amino acids, therefore creating an H-bond that holds the hand in place and prevents the conformational change [58, 59]. A study conducted by Tricoire et al. on the calcium affinity of the individual EF-hands by performing amino acid substitutions within the conserved sequence determined that the EF-I hand had a lower calcium affinity than EF-III and the deformed EF-IV [60]. The high affinity of the EF-III hand relative to the other EF-hands is believed to be directly related to the sequence of amino acid residues in EF-III, most notably their proximity to Tyr184, which contributes to the stabilization of the peroxide bound form of coelenterazine prior to light emission.

2.1.1 Engineering Low-Affinity Aequorin for Detection of Calcium in Organelles

Rational design followed by genetic engineering of aequorin to alter the binding affinity of the photoprotein to calcium can lead to mutants with calcium insensitivity. Specifically, these mutants can be created by simply replacing one of the negatively charged residues in the calcium-binding loop of an EF-hand. This type of genetic engineering and modification of aequorin can serve to increase the range of calcium detection by preparing aequorin variants with tailored binding characteristics suitable for organelles that contain a high calcium concentration. Cytosolic calcium levels are typically in the 0.5–10 μM range, suitable for wild-type aequorin, but the higher levels in the endoplasmic reticulum (ER), mitochondria, peroxisomes, and Golgi apparatus require a lower sensitivity, up to millimolar values [61]. Therefore, the investigation of the role of calcium in the function of comparatively calcium-rich organelles requires just such aequorin with a lower calcium sensitivity.

As previously mentioned, the EF-III hand has a binding loop that binds calcium with the strongest affinity (Asp117–Glu128), making this loop of particular interest to engineer low-affinity aequorin. Replacement of the highly conserved negatively charged Asp119 at the +Y position on the EF-III binding loop with the neutral amino acid alanine decreased the binding affinity of the EF-III hand by around 50-fold compared to the wild-type aequorin [62]. The calcium affinity was further reduced by introducing a second point mutation at position 28 (Asn–Leu) yielding an aequorin with a K_d in the mM range [59]. This low-affinity aequorin was developed specifically for the comparatively high levels of calcium found in the mitochondria, which would trigger the bioluminescent emission of the wild-type aequorin too prematurely for signal acquisition by the instruments used in that specific study. This low sensitivity aequorin was applied in transfected HeLa cells, and as a result of its low affinity to the calcium ions, it has reported calcium levels at millimolar concentrations during the highest levels of calcium uptake in mitochondria [63].

2.1.2 Engineering Aequorin Chimeras Targeted to Specific Regions within the Cell: Cancer and Beyond

Although aequorin with engineered affinities can be readily introduced to cells, aequorin will diffuse through the cell unless modified by an additional peptide sequence to target a specific region within the cell [25]. Without these additional modifications, the bioluminescent signal does not report calcium from only the region of interest, but also from other parts of the cell where the unmodified aequorin has diffused, generating noise. Calcium is universal in cells, however, each cell type utilizes calcium in a unique manner, requiring an understanding of the flux of calcium in each cell type and each subcellular region. For example, the calcium levels of the mitochondria are critical to maintaining healthy cell function and preventing cell death, and apoptotic and necrotic cell death is often accompanied by calcium overload [64]. Of particular interest is the role of calcium in apoptosis and cancer. Therefore, the movement of calcium between the mitochondria, the endoplasmic reticulum, which stores calcium and is associated with downstream pathophysiology when its stores of calcium are depleted, the cytosol, which uses calcium as a ubiquitous messenger, and other cellular regions have been repeatedly studied as technologies have improved [65].

A mitochondria targeted aequorin chimera with a mitochondrial presequence (mtAEQ) was first developed to demonstrate directly the relationship between agonist stimulated elevations of cytosolic calcium on mitochondrial calcium levels [66]. Transfected HeLa cells and bovine adrenal medulla chromaffin cells were both transfected with mtAEQ to provide information on the relationship between calcium concentrations and the opening of the mitochondrial permeability transition pore (PTP), the presence of which is associated with neurodegeneration and an important target for the treatment of traumatic brain injury and stroke [67, 68]. The D119A mutant aequorin (mutAEQ) developed by Montero et al. was fused to a signal sequence to target the ER. By combining mtAEQ, endoplasmic reticulum targeted aequorin (erAEQ), and cytosolic aequorin (cytAEQ), the role of the promyelocytic leukemia (PML) tumor suppressor, a nuclear protein, in apoptosis was found to be connected to calcium signaling in all three subcellular regions [69]. Another focus of targeted chimeras is to the caveolar microdomain, invaginations in the plasma membrane of some vertebrate cells that contain calcium responsive proteins such as GPCRs, receptor tyrosine kinases, and caveolin proteins. A chimera consisting of mutAEQ was fused to caveolin (Cav1–Aeq) and expressed in HeLa cells to investigate the role of the mostly cytosolic enzyme sphingosine kinase 1 (SK1), an oncogenic protein and potential predictive biomarker of cancer patient outcomes, in controlling calcium in the microdomain and its relation to cell differentiation, motility, and apoptosis [70, 71].

Further studies using mtAEQ and erAEQ have been on the role of calcium signaling in the tumor suppression mechanism of the phosphatase and tensin homologue (PTEN), that is not only linked to cancers such as prostate, breast, thyroid, and central nervous system (CNS) cancers, but possibly autism and learning disabilities [72–74]. These chimeras were utilized again to identify a new calcium uptake channel in the mitochondria dubbed the mitochondrial calcium uniporter (MCU), and further investigate the critical role of calcium movements between mitochondria and the endoplasmic reticulum in apoptosis [75–77]. The mtAEQ was used again to investigate the MCU, which can bind calcium with nanomolar affinity and is a potential therapeutic target, and the physiological role of mitochondrial calcium. Primary mouse embryonic fibroblasts (MEF) were infected with an adenovirus encoding mtAEQ to monitor changes in calcium levels following histamine stimulation, revealing that $MCU^{-}/^{-}$ knockout mice display markedly reduced calcium fluxes in the mitochondria. The lack of MCU expression was accompanied by an impairment in the ability of the knockout mice to do strenuous work and the failure of the heart to respond to the immunosuppressant cyclosporine A, which targets PTP, although not changes in the magnitude of cell death, suggesting that calcium-independent pathways emerge in the absence of the MCU [78–80].

Additional studies conducted focusing on research areas other than cancer using several targeted aequorin chimeras, cytAEQ, mtAEQ, erAEQ, and Golgi targeted aequorin (GoAEQ), revealed $MgCl_2$ could be used to restore the cytosolic and mitochondrial calcium levels in patients with the genetic dermatological condition Hailey–Hailey disease (HHD), which has symptoms attributed to a defective calcium pump [81]. An aequorin mutant fused to calsequestrin was used to target the sarcoplamic reticulum (srAEQ), the equivalent to the ER in muscle cells, and demonstrated the heterogeneous distribution of calcium in the region and the influence of calcium dysregulation on different phenotypes of malignant hyperthermia, a rare, heritable, potentially fatal condition triggered by anesthesia [82, 83]. In 2011, Giorgi et al. developed a new type of high-throughput method of studying GPCRs and tyrosine kinase receptors by combing aequorin chimeras and multiwell plate readers [84], providing a new improved method for the study of GPCRs and the development of GPCR targeting pharmaceuticals.

2.1.3 Aequorin–Fluorescent Protein Fusions for BRET and BRET-Based Multiplex Calcium Imaging in Cells

Aequorin and GFP combine to create a naturally occurring BRET reaction in the *Aequorea victoria* jellyfish, in which aequorin is the donor and GFP is the acceptor [85]. By fusing aequorin to GFP and GFP derived fluorescent proteins, a catalog of colors becomes available to provide bioluminescent imaging of calcium in multiple subcellular regions simultaneously that can be distinguished by emission wavelength [86].

Venus (YFP), RFP, mOrange, and tandem-dimer Tomato (tdTomato, tdTa) have all been fused to aequorin for applications in in vivo imaging (Fig. 4). Fusions of GFP–AEQ have been expressed in neurons incorporating a recombinant Sindbis virus for the quantitative detection of calcium transients in a single layer of neurons derived from brain slices [88], in the neurons of optically transparent transgenic

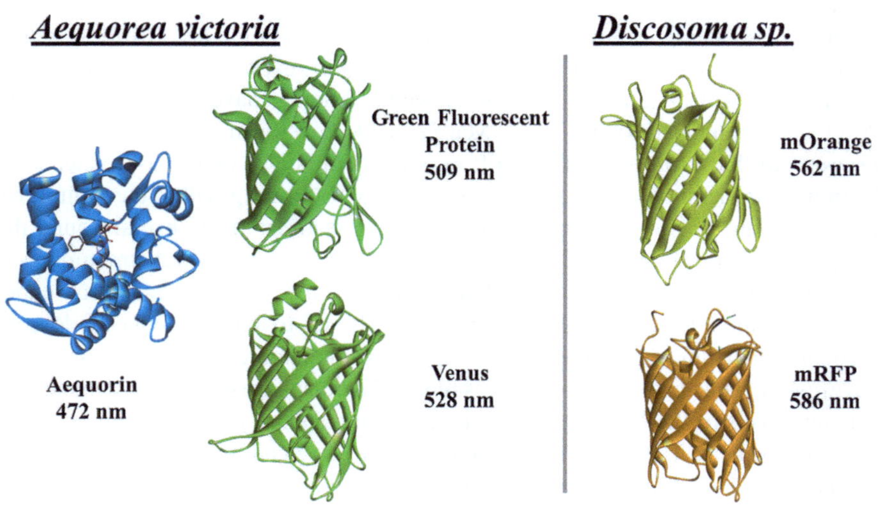

Fig. 4 Aequorin and select fluorescent proteins available for BRET. Aequorin, GFP, and the GFP derivative Venus have their origins in the Hydromedusa *Aequorea victoria*. The red fluorescent protein DsRed is native to the coral *Discosoma sp.* and is the protein from which a series of fluorescent proteins, including mOrange and mRFP, are derived. (Aequorin PDB Code: 1EJ3, GFP PDB Code: 1EMA; Venus PDB Code: 1MYW; mOrange PDB Code: 2H5O; mRFP PDB Code: 2VAD) [87]

zebrafish larva [89], and in the neurons of transgenic mice with several organelle targeted GFP–AEQ fusions [90], providing insights into communication between neurons, and their role in the larger network. Monomeric red fluorescent protein (mRFP) is a GFP mutant that can be used with aequorin for BRET and paired with AEQ–GFP for monitoring calcium in two different organelles simultaneously. By combining a GFP fusion targeting the cytosol and a mRFP fusion targeting the nucleus in human embryonic kidney (HEK-293) cells, both sections could be visualized and the signals can be differentiated by color [91]. Recently, effort has been put into the development of aequorin-based bioluminescent signals with a greater red-shift for application in deep tissue imaging. Despite the fusion of aequorin to red fluorescent proteins before, insufficient signal intensity due to poor energy transfer during BRET has prevented these chimeras from being adopted in place of other signaling molecules [92]. A recently described aequorin chimera, Redquorin, a tandem dimer with aequorin and tdTomato, displayed BRET at about 73 % of the maximum energy transfer efficiency, at 575 nm [93]. This Redquorin was applied to track transient calcium changes in HEK-293 cells and developing transfected zebrafish embryos.

2.1.4 Imaging Calcium in Whole Animals In Vivo

In comparison to calcium signaling in cells, imaging calcium in whole animals focuses calcium signaling across whole structures. Using a transgenic model, Cheung et al. were able to detect two distinct periods in the temporal characteristics of calcium signaling during the differentiation of slow muscle cells (SMCs) within forming skeletal muscles by combining aequorin expression and an α-actin promoter [94]. Calcium movements in whole animals have been tracked with AEQ–GFP fusions [95]. In 2007, Martin et al. developed a transgenic line of *Drosophila* that expressed AEQ–GFP in mushroom bodies (MBs) associated with learning and memory, and in the central complex (CC), deep within the brain [96]. By combining photon counting with continuous readings over several hours, the unique calcium response to cholinergic transmission in these neural structures was identified by whole brain imaging. Furthermore, *dunce* mutant flies, which experience impaired learning and memory, were shown to have a reduced delayed secondary calcium rise in response to nicotine stimulation compared to non-*dunce* *Drosophila.* This delayed response was reduced by thapsigargin, which depletes the calcium levels of the ER, suggesting the calcium response of the MB involves the ER compartment. Therefore, bioluminescence imaging can be used to link calcium signaling and functional mapping in the fruit fly brain. The noninvasive bioluminescence of imaging of nonanesthetized transgenic mice was also accomplished with AEQ–GFP [97]. Whole-body imaging of calcium was performed using mitochondria targeting AEQ–GFP fusion on both freely moving mice and those experiencing kainic acid-induced epileptic seizures. Additionally, whole body recordings were made of newborn mice to correlate mitochondrial calcium levels to the behavioral components of the sleep/wake cycle.

3 Engineering of Aequorin with Improved Emission Characteristics for Bioluminescent Detection and Imaging

As previously mentioned, the bioluminescence of proteins such as aequorin allows the achievement of low detection limits by eliminating the need for external excitation that can trigger fluorescence in molecules present in samples other than the analyte of interest. This makes aequorin especially well-suited to detection in samples with low concentrations of the analyte or in miniaturized analytical platforms, such as the lab-on-a-chip [6, 98], and the small volumes used in some high-throughput assays [99]. Until recently, aequorin had a significant drawback as a reporter molecule: a lack of diversity in emission color compared to fluorescent proteins. By expanding the colors of aequorin available, it can be applied in multianalyte studies such as multiplexed single-well analysis.

3.1 Rational Mutagenesis and Random Mutagenesis

Studies of the amino acid sequence of aequorin prior to the publication of the X-ray crystal structure showed the presence of the amino acids cysteine, histidine, proline, and tryptophan, which are not usually associated with calcium binding proteins [100], and these positions were subsequently selected for rational mutagenesis via substitutions with other amino acids to determine their effect on aequorin's bioluminescent activity. The global replacement of all the cysteine residues with the structurally similar serine resulted in a cysteine-free aequorin with increased luminescence [101].

Site-directed mutagenesis of aequorin focusing on histidine, proline, and tryptophan amino acids revealed the influence of these amino acids on the spectral characteristics of the protein [28, 101–105]. Replacing histidine, proline, or tryptophan with amino acids of similar properties gave insight on not only which positions affected the oxidation of coelenterazine, but provided target locations for the rational site-directed mutagenesis of aequorin for the purpose of altering its emission characteristics [106]. Once the X-ray crystal structure of aequorin was determined, it revealed a 600-Å hydrophobic pocket that provided new residues of interest that interact with coelenterazine via stabilizing hydrogen bonds or by the π–π stacking of benzene rings, in particular three separate Tyr–His–Trp triads that included some positions that had already been studied, providing further evidence of their importance to aequorin's bioluminescence emission characteristics (Fig. 5) [30].

Rational site-directed mutagenesis focused on altering the H-bonds of these triads and the bulkiness of the amino acid side-chains within the active site to alter the hydrophobic and π–π interactions of the coelenterazine so as to change the specific activity, wavelength of emission, and bioluminescent half-life without losing activity [107]. Tyrosine and phenylalanine, in particular, have been shown to

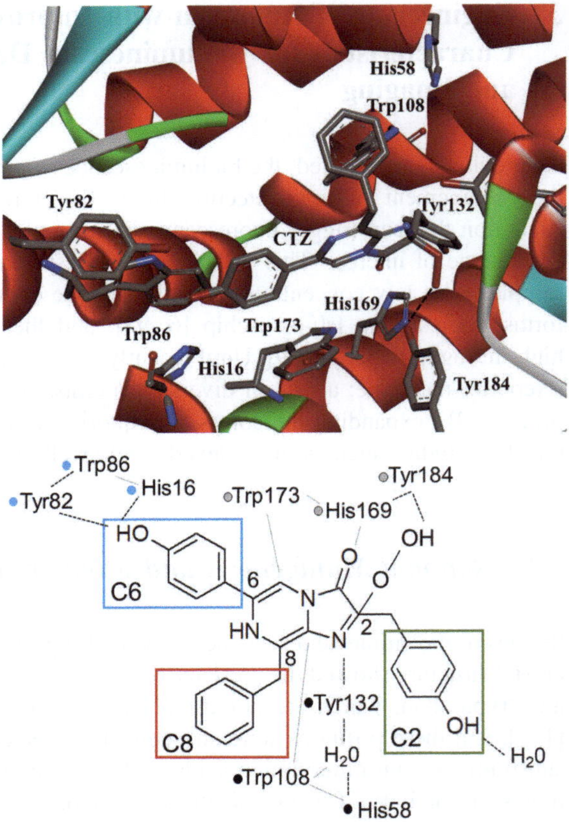

Fig. 5 The nine amino acids that form the three stabilizing Tyr–His–Trp triads are shown bracketing the hydroperoxide form of coelenterazine. The protein has been edited for clarity. *Top* The hydrogen bound between the hydroperoxide and the tyrosine at position 184 is a *dashed white line* (aequorin PDB Code: 1EJ3). *Bottom* The triads and their interactions with coelenterazine, water, and each other. Triads are grouped by *color*, the H-bonds are *dashed*, and other bonds are in *dots*

influence the alteration of the wavelength of emission. The wavelength of emission of aequorin is determined by the electronic conjugation of the π-electrons on the aromatic ring associated with the Tyr82–His16–Trp86 triad stabilizing (C6) and the central imidazopyrazinone, as determined by the torsional angle between the two rings and also the rotational freedom between the two sections along the σ-bond [30, 108, 109].

In addition to the rational design of mutants by the site-directed mutagenesis of chosen sites, random mutagenesis can be used to determine the effects of cumulative mutations not apparent ab initio. The previously aforementioned studies in Sect. 2 into the role of each amino acid on the calcium-binding EF-hands utilized random mutagenesis to generate thermostable mutants of aequorin and investigate the decay kinetics of several mutants, and confirmed a constant quantum yield with an inverse relationship and peak bioluminescent intensity [58, 110]. Furthermore, these studies suggested the decay rate of the bioluminescence of aequorin contained both fast and slow kinetics influenced by all three binding EF-hands, and that calcium binding to one hand was sufficient to trigger slow rate luminescence. In 2014, random mutagenesis was utilized to increase further the number of thermostable aequorin mutants available, including a quadruple mutant that maintained 75 % of its bioluminescent activity after 72 h at 37 °C [111].

3.2 The Incorporation of Nonnatural Amino Acids into Aequorin for the Development of Mutants for Multiplex Detection

An inherent drawback of the mutagenesis of aequorin is that alteration of the peptide sequence frequently results in considerable or even, in some cases, total loss of bioluminescence activity [107, 112]. The finite number of canonical amino acids available that can be incorporated without loss of activity has limited the number of active mutants which can be created and restricted the applications of aequorin. Attempts to expand the catalogue of aequorin has resulted in applying techniques to add nonnatural amino acids to the aequorin sequence to increase the number of amino acids past the canonical number [113]. Schultz et al. pioneered the incorporation of nonnatural amino acids into proteins by using amber codon suppression [114].

Schultz and collaborators cleverly took advantage of the fact that the TAG stop codon is never used for methionine incorporation in *Escherichia coli*, thus allowing the TAG codon to be used for the incorporation of nonnatural amino acids through "amber suppression". By creating a transfer RNA (tRNA) and transfer tRNA synthetase (tRNAse) that charges a specific nonnatural amino acid and targets the TAG codon during translation, that nonnatural amino can be incorporated such that the resultant protein either has incorporated the chosen nonnatural amino acid or is truncated when the TAG is read as a stop codon in its absence [115–117]. By transforming a cell line with a plasmid containing the tRNA/tRNAse and a plasmid containing the aequorin with TAG codon at the position for the incorporation of the nonnatural amino acid, we were able to prepare several aequorin mutants containing nonnatural amino acids [112].

The tyrosine at position 82 that stabilizes (C6) in cysteine-free aequorin has been substituted with four nonnatural amino acids *p*-aminophenylalanine, *p*-bromophenylalanine, *p*-iodophenylalanine, and *p*-methoxyphenylalanine, which were then combined with various coelenterazine analogues (Fig. 6) [112]. Each of the mutants displayed altered spectral characteristics compared to wild-type aequorin, creating a catalogue of semi-synthetic aequorins with a range of peak emission wavelengths spanning 72 nm. The red-shifts in the wavelength of the peak bioluminescence emission of the aequorin incorporating *p*-aminophenylalanine and *p*-methoxyphenylalanine were consistently of 8–10 nm independent of the coelenterazine analogue incorporated.

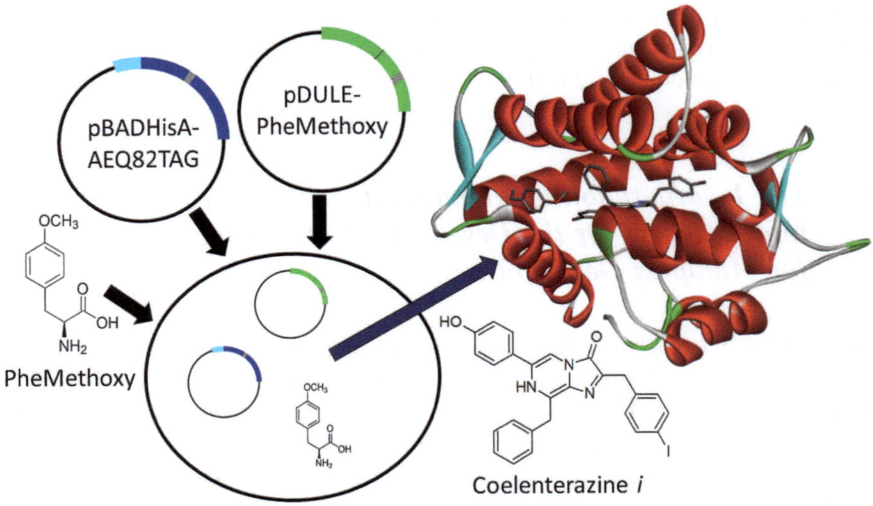

Fig. 6 Schematic of the incorporation of nonnatural amino acids into aequorin and subsequent pairing with a coelenterazine analogue. The phenylalanine analogue, *p*-methoxyphenylalanine (PheMethoxy) was incorporated into aequorin (*indigo*) using a TAG codon (*gray*). A His6x TAG was included at the N-terminus for purification (*turquoise*). The pDULE contains a unique *p*-methoxyphenylalanine targeting tRNA/tRNAse that incorporates the nonnatural amino acid at the UAG codon in RNA (aequorin with coelenterazine *i* PDB Code: 1UHI)

3.3 Applications of Engineered Aequorin in In Vitro Assays for Clinical, Pharmaceutical, and Environmental Detection and Analysis

Aequorin, including the cysteine-free aequorin, has been used in a large number of assays, including binding and whole-cell assays due to the natural low detection limits of aequorin and safety and ease of use compared to alternative labels such as radioactive labels [118, 119]. The increased versatility of aequorin made possible by recent genetic engineering has increased the number of aequorin-based assays even further to cover a range of applications in clinical, pharmaceutical, and environmental detection and analysis. Target genes, proteins, metabolic small molecules, and environmental pollutants are all focuses of aequorin-based bioluminescent assays.

3.3.1 Aequorin-Based Molecular Switches and Applications in the Detection of Clinically Relevant Molecules

The genetic manipulation of aequorin expands beyond point mutations to include splitting the aequorin. Aequorin can be split into two fragments, causing total activity loss, then be reassembled to regain its bioluminescent activity [120]. This reassembly

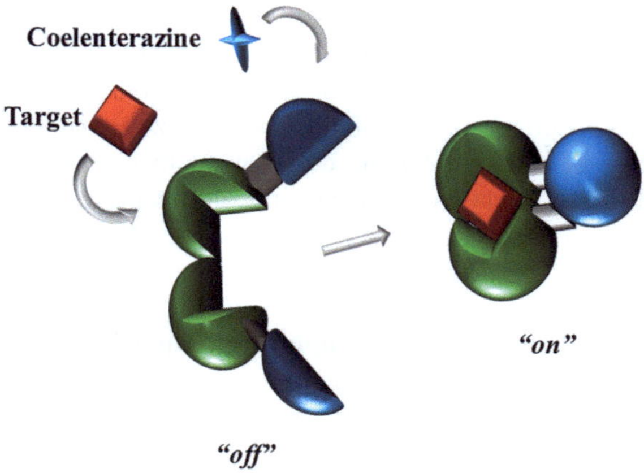

Fig. 7 Schematic of a molecular switch demonstrating its "*on/off*" states. The molecular switch was prepared by genetically splicing aequorin into two fragments and inserting the gene of a hinge-motion binding protein in between the spliced aequorin. In the absence of the ligand, the spliced aequorin cannot reassemble to emit light. Upon binding to its ligand, the hinge-motion binding protein changes conformation, bringing together the two fragments of aequorin, and thus, allowing the aequorin to reassemble and emit light in the presence of coelenterazine and calcium

is facilitated either by spontaneous assembly, close contact, or with the aid of additional protein or proteins. Molecular switches are fusion proteins that reassemble the fragments by means of a binding protein that, upon binding of the analyte, undergoes a conformational change that brings the two halves together [121–123]. These two orientations are referred to as "off" and "on" (Fig. 7). Molecular switches can also reverse the "on" and "off" positions, so that luminescence is present without the analyte, but absent when the analyte is bound. Although most genetically fused aequorin and recognition elements are fused in tandem at the aequorins N or C-terminus, molecular switches insert the recognition protein between the two fragments of the split aequorin. Upon the binding of the analyte by the recognition protein the two separated aequorin fragments are brought together to restore the signal, resulting in "on/off" protein conformations. The opposite orientations, with the aequorin fragments together in the "on" position in the absence of the analyte and separated in the "off" position during binding, are also possible. Aequorin-based molecular switches include detectors for glucose using glucose binding protein, important secondary messenger cAMP, and sulfates, a small ionic molecule linked to autism [120, 124, 125].

3.3.2 Aequorin in Homogeneous and Competitive Hydrophobic Assays for Clinical and Environmental Applications

As with molecular switches, homogeneous assays do not require the immobilization and subsequent incubation and washing steps of the DNA and immunoassays. The process is streamlined by the presence of the analyte itself causing a detectable change, such as effecting the signal intensity. Homogeneous assays using aequorin include those detecting the activation of GPCRs [42], biotin, and cortisol [126, 127]. In 2012, the Daunert lab produced a bioluminescence inhibition assay based on the competitive binding of the coelenterazine and the analyte in the hydrophobic core of aequorin for the detection of hydroxylated polychlorinated biphenyls (OH-PCBs), a class of contaminants and suspected endocrine disruptor [128].

3.3.3 DNA Hybridization Probes Employing Aequorin as the Label for Clinical Diagnostics

Another application is the sensitive and rapid detection of DNA associated with illnesses such as cancer [129], and pathogens, including HIV and *Chlamydia* [130–132]. DNA hybridization assays utilizing aequorin are able to detect target DNA down to 0.25 attomoles [133]. A DNA hybridization assay for the detection of the deadliest form of malaria, *Plasmodium falciparum*, was able to detect down to 3 pg/μL, and could be adapted to multiplex analysis [134]. A quantitative PCR (qPCR) method for the diagnosis of another pathogen, the fungus *Aspergillus*, the cause of invasive pulmonary aspergillosis (IPA) used aequorin as an internal amplification control (IAC). The assay was able to detect *Aspergillus* DNA at a qPCR cutoff of 13 fg of *Aspergillus* genomic DNA and was able to detect the DNA in 76.9 % of subjects showing proven or probable IPA [135]. DNA and RNA hybridization assays are based on a labeled DNA or RNA sequence complementary to another immobilized sequence. Aequorin-based detection can be combined with microfluidic platforms to achieve the detection of single nucleotide polymorphism (SNPs) at the attomolar level in real-time [136]. A dual-analyte bio/chemiluminescent method for the detection of both the normal and mutant allele products of the primer extension reaction (PEXT) was developed by Christopoulos et al. [137, 138]. Primers containing mutant deoxyuridine triphosphate (dUTP) labeled with biotin and digoxigenin are added to PCR-amplified DNA fragments to genotype the section of interest to begin the detection of SNPs. The reaction products are captured using microtiter well-plates containing immobilized oligo (dT) strands, then incubated with streptavidin–aequorin and antidigoxigenin–alkaline phosphatase (ALP) conjugates for bioluminescent and chemiluminescent detection. This method proved effective for quantitative PCR, digoxigenin and prostate-specific antigens, and transgenes and SNPs. High-throughput methods for SNP detection using aequorin have also been developed to look for a pair of gene mutations in the toll-like receptor (TLR)-4 gene, to aid in the study of the role of these SNPs in their associated diseases [139].

3.3.4 Recent Applications of Aequorin in Immunoassays for the Detection and Quantitative Analysis of Clinically Relevant Molecules

Immunoassays are sensitive binding assays that use antibodies or other recognition peptides as the recognition elements for the analyte. When aequorin is used in conjunction with these recognition elements using either chemical conjugation or genetic fusion highly sensitive assays can be developed. Aequorin has been genetically fused to fab fragments while maintaining both the binding affinity of the antibody and the bioluminescence of aequorin for sensitive assays [140–142]. It can also be chemically conjugated to produce aequorin–antibody pairs for assays for the detection of several biologically relevant molecules down to attomole levels. Aequorin is a common reporter in immunoassays and has been covered in reviews before, therefore, this section only covers the most significant as well as the most recent developments [143–145].

Aequorin can be combined with other reporters, such as luciferase, to create assays with more than one type of reporter [146]. Assays for the dual detection of prostatic acid phosphatase (PAP), prostate specific antigen (PSA), and α-fetoprotein (AFP), were developed with aequorin-linked antibody fragments showing minimum detection limits of 9.4×10^{-21} mol per assay [147]. Alternatively, aequorin can be linked to the analyte instead. In the second approach, the labeled analyte then competes with the unlabeled analyte present in the sample for the limited number of antibody binding sites available on one of the many optional platforms (e.g., microtiter plates, polystyrene beads, magnetic microsphere, and paper strips). ELISA-type assays of this cover a range of biologically significant molecules, such as neurotransmitters, protein c, noted stress-related molecule cortisol, and cardiac-relevant molecules in a dual assay detecting angiotensin II and 6-keto-prostaglandin-F1-alpha using aequorin with two different coelenterazine analogues, providing detection limits as low as 1 pg/mL [20, 22, 148–150]. A third approach is to use an intermediary molecule to connect the antibody and aequorin. For example, aequorin and the selected antibody can be randomly or site-specifically modified to allow for linking to molecules that contain a streptavidin or a biotin, providing a means of linking the two proteins by a streptavidin–biotin bond. Immunoassays that use biotinylated aequorin have been used to detect analytes at levels as low as 0.02 ng/mL [144, 151, 152].

3.3.5 Whole-Cell Assays Employing Aequorin as a Reporter for the Detection of Pathogens, Antibodies, and the Detection and Analysis of the Toxicity of Environmental Pollutants

Unlike other systems, whole-cell assays, or whole-cell biosensors, only report on bioavailable portions of a target sample, making them especially suited to the detection of pollutants and quantification of the toxic cumulative effect that pollutants have on living organisms. Whole-cell assays often link the expression of aequorin to the presence of a target analyte that enters the cell [153, 154]. Fungal lines can also be used to express aequorin in response to analytes. A short-term toxicity whole-cell assay using *Aspergillus awamori* was designed to sense chromium and zinc ions and the phenolic polar narcotic 3,5-DCP [155]. Whole-cell sensors have also been developed for detection of pollutants such as toluene and related compounds [156].

For detection of biological molecules, Rider et al. created a programmable form of whole-cell biosensors by using B lymphocytes. Pathogen-specific B cell lines generated from M12g3R (IgM+) B cell lines that consistently express cytosolic aequorin were transfected with plasmids containing antibodies with variable regions specific for the pathogens [157]. These whole-cell biosensors were able to discriminate between pathogenic and nonpathogenic *E. coli* strains. A more recent assay using aequorin for the detection of biological analytes was developed by Araki et al. for the detection of thyroid-stimulating antibodies (TSAb), the cause of hyperthyroidism in the autoimmune disorder Graves' disease. The TSAb aequorin assay can be conducted in 4 h without sterile conditions and was positive in 98.9 % of untreated Graves' disease patients, making it more sensitive than conventional assays [158].

Cells containing the assays can be immobilized on platforms such as optical fibers or encapsulated into sol-gels, or used with microfluidic platforms to create high-throughput arrays [159]. The reagents of whole-cell assays can also be stored for the long term using methods as simple as freezing, and more complex techniques such as encapsulation within biocompatible polymers and sporulation [160–163].

4 In Vivo Detection and Imaging with Aequorin and Aequorin Chimeras

In vivo detection has even greater challenges than in vitro studies using biological samples. Biological tissue has an optical window that narrows the range of wavelengths available for reporters to between 650 and 950 nm [164, 165], and tissue penetration depth is limited to between 1 and 2 cm [166]. Although several options exist that do not require a luminescent signal, such as radiolabels, near-infrared quantum dots, and near-infrared fluorescing carbon nanotubes, all of these have the risk of toxicity to living organisms, and in the case of carbon nanotubes, damage even from accidental inhalation [167–169].

In vivo detection with aequorin has been applied only in a limited capacity due to the problems associated with native aequorin emission wavelength in the blue region. As mentioned in Sect. 2, aequorin has recently been paired with fluorescent proteins with red emission wavelengths capable of being transmitted through the blood with minimal absorption. This opens the door for applications of aequorin in deep tissue in vivo by addressing the signal absorption currently preventing broader applications of aequorin in this area of detection [170]. Some examples of using aequorin chimeras such as using aequorin conjugated to a designed ankyrin repeat protein (DARPin) targeting vascular endothelial growth factor A has been used for imaging age-related macular degeneration in mice [171].

5 Conclusions and Future Perspectives

Although the broad application of aequorin does not have the same long history of other luminescent proteins such as GFP, the impact of aequorin is being felt as it finds increasing use in the field of biotechnology. Aequorin's affinity for calcium, tunable emission characteristics, ability to function as a bioluminescent label at ultralow concentrations and volumes, temperature stability, and adaptability towards its incorporation into different analytical platforms has made it a valuable tool in a variety of fields, not just in research, but also in translational and clinical medicine.

Calcium is an important part of intracellular signaling, and calcium signaling pathways have a wide variety of properties across cell types that require their own reviews to cover in detail [34]. Among other roles, calcium plays a key role in regulating cell growth and apoptosis and the dysregulation of calcium is linked to several cancers and is of concern in a number of illnesses and disorders, from dermatological conditions to Alzheimer's [172, 173]. Moreover, calcium is the second messenger released by GPCRs, the target of over 50 % of all pharmaceuticals, and assays used for high-throughput screening of GPCR-targeting pharmaceuticals use aequorin [43]. The highly sensitive calcium detection and tracking provided by aequorin has been a key part of the investigation of calcium in both these areas, providing continuing advancements in both clinical research and pharmaceutical development.

The bioluminescent signal of aequorin can also be used to detect and quantify a wide variety of small molecules, proteins, and genes, by linking the aequorin to the targeting molecule. The versatility of aequorin covers applications in clinical, pharmaceutical, and environmental assays, and is increasingly being exploited for use in imaging. Aequorin does not suffer from the background fluorescence of GFP, the special handling and disposal of radio labels, or the lack of imaging at the molecular level of MRI, and as red-shifted aequorin becomes available, the applications of aequorin in deep tissue will become broader.

Aequorin has already been used in assays that take a variety of approaches to the detection and quantification of analytes in vitro and been successfully utilized for in vivo imaging and the real-time monitoring of cellular processes. As new techniques emerge for the genetic engineering of proteins as well as a more complete understanding of the role of each part of the complexed protein in its sensitivity and emission characteristics, aequorin will reach its full potential as a photoprotein for cellular signal detection.

References

1. Eglen RM, Reisine T (2008) Photoproteins: important new tools in drug discovery. Assay Drug Dev Technol 6(5):659–671. doi:10.1089/adt.2008.160
2. Rowe L, Dikici E, Daunert S (2009) Engineering bioluminescent proteins: expanding their analytical potential. Anal Chem 81(21):8662–8668. doi:10.1021/Ac9007286
3. The Royal Swedish Academy of Sciences (2008) The Nobel Prize in Chemistry 2008-Press Release
4. Scott D, Dikici E, Ensor M, Daunert S (2011) Bioluminescence and its impact on bioanalysis. Annu Rev Anal Chem (Palo Alto Calif) 4:297–319. doi:10.1146/annurev-anchem-061010-113855
5. Shimomura O, Johnson FH, Saiga Y (1962) Extraction, purification and properties of aequorin, a bioluminescent protein from the luminous hydromedusan, Aequorea. J Cell Compar Physiol 59:223–239
6. Lewis JC, Daunert S (2000) Photoproteins as luminescent labels in binding assays. Fresen J Anal Chem 366(6–7):760–768
7. Roda A, Pasini P, Mirasoli M, Michelini E, Guardigli M (2004) Biotechnological applications of bioluminescence and chemiluminescence. Trends Biotechnol 22(6):295–303. doi:10.1016/j.tibtech.2004.03.011
8. Prendergast FG (2000) Bioluminescence illuminated. Nature 405(6784):291–293. doi:10.1038/35012734
9. Shimomura O, Johnson FH (1975) Chemical nature of bioluminescence systems in coelenterates. Proc Natl Acad Sci USA 72(4):1546–1549
10. Shimomura O, Inoue S, Johnson FH, Haneda Y (1980) Widespread occurrence of coelenterazine in marine bioluminescence. Comp Biochem Phys B 65(2):435–437. doi:10.1016/0305-0491(80)90044-9
11. Nakatsu T, Ichiyama S, Hiratake J, Saldanha A, Kobashi N, Sakata K, Kato H (2006) Structural basis for the spectral difference in luciferase bioluminescence. Nature 440(7082):372–376. doi:10.1038/nature04542
12. Marques SM, da Silva JCGE (2009) Firefly bioluminescence: a mechanistic approach of luciferase catalyzed reactions. IUBMB Life 61(1):6–17. doi:10.1002/Iub.134
13. Widder EA (2010) Bioluminescence in the Ocean: Science 328(5979):704–708. doi:10.1126/science.1174269
14. Haddock SH, Moline MA, Case JF (2010) Bioluminescence in the sea. Ann Rev Mar Sci 2:443–493. doi:10.1146/annurev-marine-120308-081028
15. Chen SF, Ferre N, Liu YJ (2013) QM/MM study on the light emitters of aequorin chemiluminescence, bioluminescence, and fluorescence: a general understanding of the bioluminescence of several marine organisms. Chemistry 19(26):8466–8472. doi:10.1002/chem.201300678
16. Brini M (2008) Calcium-sensitive photoproteins. Methods 46(3):160–166. doi:10.1016/j.ymeth.2008.09.011
17. van Oort B, Eremeeva EV, Koehorst RBM, Laptenok SP, van Amerongen H, van Berkel WJH, Malikova NP, Markova SV, Vysotski ES, Visser AJWG, Lee J (2009) Picosecond fluorescence relaxation spectroscopy of the calcium-discharged photoproteins aequorin and obelin. Biochem Us 48(44):10486–10491. doi:10.1021/Bi901436m
18. Inouye S, Sato J, Sahara-Miura Y (2011) Recombinant Gaussia luciferase with a reactive cysteine residue for chemical conjugation: expression, purification and its application for bioluminescent immunoassays. Biochem Biophys Res Commun 410(4):792–797. doi:10.1016/j.bbrc.2011.06.063
19. Malikova NP, Burakova LP, Markova SV, Vysotski ES (2014) Characterization of hydromedusan Ca(2+)-regulated photoproteins as a tool for measurement of Ca(2+) concentration. Anal Bioanal Chem 406(23):5715–5726. doi:10.1007/s00216-014-7986-2

20. Lewis JC, Cullen LC, Daunert S (2000) Site-specifically labeled photoprotein-thyroxine conjugates using aequorin mutants containing unique cysteine residues: applications for binding assays (Part II). Bioconjug Chem 11(2):140–145
21. Deo SK, Daunert S (2001) An immunoassay for Leu-enkephalin based on a C-terminal aequorin-peptide fusion. Anal Chem 73(8):1903–1908
22. Chiesa A, Rapizzi E, Tosello V, Pinton P, de Virgilio M, Fogarty KE, Rizzuto R (2001) Recombinant aequorin and green fluorescent protein as valuable tools in the study of cell signalling. Biochem J 355(Pt 1):1–12
23. Mirasoli M, Deo SK, Lewis JC, Roda A, Daunert S (2002) Bioluminescence immunoassay for cortisol using recombinant aequorin as a label. Anal Biochem 306(2):204–211
24. Dupriez VJ, Maes K, Le Poul E, Burgeon E, Detheux M (2002) Aequorin-based functional assays for G-protein-coupled receptors, ion channels, and tyrosine kinase receptors. Receptors Channels 8(5–6):319–330
25. Eisenstein M (2009) GPCRs: insane in the membrane. Nat Methods 6(12):929–933. doi:10.1038/nmeth1209-929
26. Bonora M, Giorgi C, Bononi A, Marchi S, Patergnani S, Rimessi A, Rizzuto R, Pinton P (2013) Subcellular calcium measurements in mammalian cells using jellyfish photoprotein aequorin-based probes. Nat Protoc 8(11):2105–2118. doi:10.1038/nprot.2013.127
27. Kawasaki H, Nakayama S, Kretsinger RH (1998) Classification and evolution of EF-hand proteins. Biometals 11(4):277–295. doi:10.1023/A:1009282307967
28. Lewit-Bentley A, Rety S (2000) EF-hand calcium-binding proteins. Curr Opin Struct Biol 10 (6):637–643
29. Eremeeva EV, Markova SV, Westphal AH, Visser AJWG, van Berkel WJH, Vysotski ES (2009) The intrinsic fluorescence of apo-obelin and apo-aequorin and use of its quenching to characterize coelenterazine binding. FEBS Lett 583(12):1939–1944. doi:10.1016/j.febslet.2009.04.043
30. Shimomura O, Johnson FH (1978) Peroxidized coelenterazine, the active group in the photoprotein aequorin. Proc Natl Acad Sci USA 75(6):2611–2615
31. Head JF, Inouye S, Teranishi K, Shimomura O (2000) The crystal structure of the photoprotein aequorin at 2.3 A resolution. Nature 405(6784):372–376. doi:10.1038/35012659
32. Shimomura O, Johnson FH (1973) Chemical nature of the light emitt er in bioluminescence of aequorin. Tetrahedron Lett 31:2963–2966
33. Miller AL, Karplus, E., Jaffe, LF (1994) Imaging [Ca^{2+}] with aequorin using a photon imaging detector. In: Nuccitelli R (ed) Methods in cell biology, vol 40. Academic Press, San Diego
34. Webb SE, Karplus E, Miller AL (2013) Retrospective on the development of aequorin and aequorin-based imaging to visualize changes in intracellular free [Ca]. Mol Reprod Dev. doi:10.1002/mrd.22298
35. Berridge MJ, Lipp P, Bootman MD (2000) The versatility and universality of calcium signalling. Nat Rev Mol Cell Biol 1(1):11–21. doi:10.1038/35036035
36. Webb SE, Fluck RA, Miller AL (2011) Calcium signaling during the early development of medaka and zebrafish. Biochimie 93(12):2112–2125. doi:10.1016/j.biochi.2011.06.011
37. Giorgi C, Baldassari F, Bononi A, Bonora M, De Marchi E, Marchi S, Missiroli S, Patergnani S, Rimessi A, Suski JM, Wieckowski MR, Pinton P (2012) Mitochondrial Ca(2+) and apoptosis. Cell Calcium 52(1):36–43. doi:10.1016/j.ceca.2012.02.008
38. Brini M, Ottolini D, Cali T, Carafoli E (2013) Calcium in health and disease. Metal Ions Life Sci 13:81–137. doi:10.1007/978-94-007-7500-8_4
39. Filmore D (2004) It's a GCPR world. Mod Drug Discov (ACS) 7(11):24–28
40. Miller KJ, Murphy BJ, Pelleymounter MA (2004) Central G-Protein Coupled Receptors (GPCR)s as molecular targets for the treatment of obesity: assets, liabilities and development status. Curr Drug Targets CNS Neurol Disord 3(5):357–377
41. Shimomura O, Johnson FH, Saiga Y (1963) Microdetermination of calcium by Aequorin luminescence. Science 140(3573):1339–1340. doi:10.1126/science.140.3573.1339

42. Le Poul E, Hisada S, Mizuguchi Y, Dupriez VJ, Burgeon E, Detheux M (2002) Adaptation of aequorin functional assay to high throughput screening. J Biomol Screen 7(1):57–65. doi:10.1089/108705702753520341
43. Klabunde T, Hessler G (2002) Drug design strategies for targeting G-protein-coupled receptors. Chembiochem: A Eur J Chem Biol 3(10):928–944. doi:10.1002/1439-7633(20021004)3:10<928:AID-CBIC928>3.0.CO;2-5
44. Haq N, Grose D, Ward E, Chiu O, Tigue N, Dowell SJ, Powell AJ, Chen MX (2013) A high-throughput assay for connexin 43 (Cx43, GJA1) gap junctions using codon-optimized aequorin. Assay Drug Dev Technol 11(2):93–100. doi:10.1089/adt.2012.469
45. Pozzan T, Rudolf R (2009) Measurements of mitochondrial calcium in vivo. Biochim Biophys Acta 1787(11):1317–1323. doi:10.1016/j.bbabio.2008.11.012
46. Davies SA, Terhzaz S (2009) Organellar calcium signalling mechanisms in Drosophila epithelial function. J Experiment Biol 212(Pt 3):387–400. doi:10.1242/jeb.024513
47. Berridge MJ, Bootman MD, Roderick HL (2003) Calcium signalling: dynamics, homeostasis and remodelling. Nat Rev Mol Cell Biol 4(7):517–529. doi:10.1038/nrm1155
48. Rizzuto R, Pozzan T (2006) Microdomains of intracellular Ca^{2+}: molecular determinants and functional consequences. Physiol Rev 86(1):369–408. doi:10.1152/physrev.00004.2005
49. Mellstrom B, Savignac M, Gomez-Villafuertes R, Naranjo JR (2008) Ca^{2+}-operated transcriptional networks: molecular mechanisms and in vivo models. Physiol Rev 88 (2):421–449. doi:10.1152/physrev.00041.2005
50. Grienberger C, Konnerth A (2012) Imaging calcium in neurons. Neuron 73(5):862–885. doi:10.1016/j.neuron.2012.02.011
51. Bootman MD (2012) Calcium signaling. Cold Spring Harb Perspect Biol 4(7):a011171. doi:10.1101/cshperspect.a011171
52. Ohashi W, Inouye S, Yamazaki T, Hirota H (2005) NMR analysis of the Mg^{2+}-binding properties of aequorin, a Ca^{2+}-binding photoprotein. J Biochem 138(5):613–620. doi:10.1093/jb/mvi164
53. Grabarek Z (2006) Structural basis for diversity of the EF-hand calcium-binding proteins. J Mol Biol 359(3):509–525. doi:10.1016/j.jmb.2006.03.066
54. Dudev T, Lim C (2003) Principles governing Mg, Ca, and Zn binding and selectivity in proteins. Chem Rev 103(3):773–788. doi:10.1021/cr020467n
55. Ohashi W, Inouye S, Yamazaki T, Doi-Katayama Y, Yokoyama S, Hirota H (2005) Backbone 1H, 13C and 15N resonance assignments for the Mg^{2+}-bound form of the Ca^{2+}-binding photoprotein aequorin. J Biomol NMR 31(4):375–376. doi:10.1007/s10858-005-1609-3
56. Biagioli M, Pinton P, Scudiero R, Ragghianti M, Bucci S, Rizzuto R (2005) Aequorin chimeras as valuable tool in the measurement of Ca^{2+} concentration during cadmium injury. Toxicology 208(3):389–398. doi:10.1016/j.tox.2004.11.038
57. Strynadka NC, James MN (1989) Crystal structures of the helix-loop-helix calcium-binding proteins. Annu Rev Biochem 58:951–998. doi:10.1146/annurev.bi.58.070189.004511
58. Tricoire L, Tsuzuki K, Courjean O et al (2006) Calcium dependence of aequorin bioluminescnece dissected by random mutagenesis. cPNAS 103(25):5
59. de la Fuente S, Fonteriz RI, de la Cruz PJ, Montero M, Alvarez J (2012) Mitochondrial free [Ca(2+)] dynamics measured with a novel low-Ca(2+) affinity aequorin probe. Biochem J 445(3):371–376. doi:10.1042/BJ20120423
60. Tricoire L, Tsuzuki K, Courjean O, Gibelin N, Bourout G, Rossier J, Lambolez B (2006) Calcium dependence of aequorin bioluminescence dissected by random mutagenesis. Proc Natl Acad Sci USA 103(25):9500–9505. doi:10.1073/pnas.0603176103
61. Villalobos C, Alonso MT, Garcia-Sancho J (2009) Bioluminescence imaging of calcium oscillations inside intreacellular organelles In: Rich PB, Doulliet C (ed) Bioluminescence. Methods in molecular biology, vol 574. Humana Press, New York, pp 203–214
62. Montero M, Brini M, Marsault R, Alvarez J, Sitia R, Pozzan T, Rizzuto R (1995) Monitoring dynamic changes in free Ca^{2+} concentration in the endoplasmic reticulum of intact cells. EMBO J 14(22):5467–5475

63. de la Fuente S, Matesanz-Isabel J, Fonteriz RI, Montero M, Alvarez J (2014) Dynamics of mitochondrial Ca^{2+} uptake in MICU1-knockdown cells. Biochem J 458(1):33–40. doi:10.1042/BJ20131025
64. Orrenius S, Zhivotovsky B, Nicotera P (2003) Regulation of cell death: the calcium-apoptosis link. Nat Rev Mol Cell Biol 4(7):552–565. doi:10.1038/nrm1150
65. Williams GS, Boyman L, Chikando AC, Khairallah RJ, Lederer WJ (2013) Mitochondrial calcium uptake. Proc Natl Acad Sci USA 110(26):10479–10486. doi:10.1073/pnas.1300410110
66. Rizzuto R, Simpson AW, Brini M, Pozzan T (1992) Rapid changes of mitochondrial Ca^{2+} revealed by specifically targeted recombinant aequorin. Nature 358(6384):325–327. doi:10.1038/358325a0
67. Montero M, Alonso MT, Carnicero E, Cuchillo-Ibanez I, Albillos A, Garcia AG, Garcia-Sancho J, Alvarez J (2000) Chromaffin-cell stimulation triggers fast millimolar mitochondrial Ca^{2+} transients that modulate secretion. Nat Cell Biol 2(2):57–61. doi:10.1038/35000001
68. Vay L, Hernandez-SanMiguel E, Lobaton CD, Moreno A, Montero M, Alvarez J (2009) Mitochondrial free [Ca^{2+}] levels and the permeability transition. Cell Calcium 45(3):243–250. doi:10.1016/j.ceca.2008.10.007
69. Giorgi C, Ito K, Lin HK, Santangelo C, Wieckowski MR, Lebiedzinska M, Bononi A, Bonora M, Duszynski J, Bernardi R, Rizzuto R, Tacchetti C, Pinton P, Pandolfi PP (2010) PML regulates apoptosis at endoplasmic reticulum by modulating calcium release. Science 330(6008):1247–1251. doi:10.1126/science.1189157
70. Pulli I, Blom T, Lof C, Magnusson M, Rimessi A, Pinton P, Tornquist K (2015) A novel chimeric aequorin fused with caveolin-1 reveals a sphingosine kinase 1-regulated Ca microdomain in the caveolar compartment. Biochim Biophys Acta. doi:10.1016/j.bbamcr.2015.04.005
71. Zhang Y, Wang Y, Wan Z, Liu S, Cao Y, Zeng Z (2014) Sphingosine kinase 1 and cancer: a systematic review and meta-analysis. PLoS ONE 9(2):e90362. doi:10.1371/journal.pone.0090362
72. Hollander MC, Blumenthal GM, Dennis PA (2011) PTEN loss in the continuum of common cancers, rare syndromes and mouse models. Nat Rev Cancer 11(4):289–301. doi:10.1038/nrc3037
73. Napoli E, Ross-Inta C, Wong S, Hung C, Fujisawa Y, Sakaguchi D, Angelastro J, Omanska-Klusek A, Schoenfeld R, Giulivi C (2012) Mitochondrial dysfunction in Pten haplo-insufficient mice with social deficits and repetitive behavior: interplay between Pten and p53. PLoS ONE 7(8):e42504. doi:10.1371/journal.pone.0042504
74. Bononi A, Bonora M, Marchi S, Missiroli S, Poletti F, Giorgi C, Pandolfi PP, Pinton P (2013) Identification of PTEN at the ER and MAMs and its regulation of Ca(2+) signaling and apoptosis in a protein phosphatase-dependent manner. Cell Death Differ 20(12):1631–1643. doi:10.1038/cdd.2013.77
75. De Stefani D, Raffaello A, Teardo E, Szabo I, Rizzuto R (2011) A forty-kilodalton protein of the inner membrane is the mitochondrial calcium uniporter. Nature 476(7360):336–340. doi:10.1038/nature10230
76. De Stefani D, Bononi A, Romagnoli A, Messina A, De Pinto V, Pinton P, Rizzuto R (2012) VDAC1 selectively transfers apoptotic Ca^{2+} signals to mitochondria. Cell Death Differ 19(2):267–273. doi:10.1038/cdd.2011.92
77. Giorgi C, Bonora M, Sorrentino G, Missiroli S, Poletti F, Suski JM, Galindo Ramirez F, Rizzuto R, Di Virgilio F, Zito E, Pandolfi PP, Wieckowski MR, Mammano F, Del Sal G, Pinton P (2015) p53 at the endoplasmic reticulum regulates apoptosis in a Ca^{2+} -dependent manner. Proc Natl Acad Sci USA 112(6):1779–1784. doi:10.1073/pnas.1410723112
78. Pan X, Liu J, Nguyen T, Liu C, Sun J, Teng Y, Fergusson MM, Rovira II, Allen M, Springer DA, Aponte AM, Gucek M, Balaban RS, Murphy E, Finkel T (2013) The physiological role of mitochondrial calcium revealed by mice lacking the mitochondrial calcium uniporter. Nat Cell Biol 15(12):1464–1472. doi:10.1038/ncb2868

79. Patron M, Raffaello A, Granatiero V, Tosatto A, Merli G, De Stefani D, Wright L, Pallafacchina G, Terrin A, Mammucari C, Rizzuto R (2013) The mitochondrial calcium uniporter (MCU): molecular identity and physiological roles. J Biol Chem 288(15):10750–10758. doi:10.1074/jbc.R112.420752
80. Qiu J, Tan YW, Hagenston AM, Martel MA, Kneisel N, Skehel PA, Wyllie DJ, Bading H, Hardingham GE (2013) Mitochondrial calcium uniporter Mcu controls excitotoxicity and is transcriptionally repressed by neuroprotective nuclear calcium signals. Nat Commun 4:2034. doi:10.1038/ncomms3034
81. Borghi A, Rimessi A, Minghetti S, Corazza M, Pinton P, Virgili A (2015) Efficacy of magnesium chloride in the treatment of Hailey-Hailey disease: from serendipity to evidence of its effect on intracellular Ca(2+) homeostasis. Int J Dermatol 54(5):543–548. doi:10.1111/ijd.12410
82. Brini M, De Giorgi F, Murgia M, Marsault R, Massimino ML, Cantini M, Rizzuto R, Pozzan T (1997) Subcellular analysis of Ca^{2+} homeostasis in primary cultures of skeletal muscle myotubes. Mol Biol Cell 8(1):129–143
83. Brini M, Manni S, Pierobon N, Du GG, Sharma P, MacLennan DH, Carafoli E (2005) Ca^{2+} signaling in HEK-293 and skeletal muscle cells expressing recombinant ryanodine receptors harboring malignant hyperthermia and central core disease mutations. J Biol Chem 280 (15):15380–15389. doi:10.1074/jbc.M410421200
84. Giorgi C, Romagnoli A, Agnoletto C, Bergamelli L, Sorrentino G, Brini M, Pozzan T, Meldolesi J, Pinton P, Rizzuto R (2011) Translocation of signalling proteins to the plasma membrane revealed by a new bioluminescent procedure. BMC Cell Biol 12:27. doi:10.1186/1471-2121-12-27
85. Roda A, Guardigli M, Michelini E, Mirasoli M (2009) Nanobioanalytical luminescence: Forster-type energy transfer methods. Anal Bioanal Chem 393(1):109–123. doi:10.1007/s00216-008-2435-8
86. Rogers KL, Stinnakre J, Agulhon C, Jublot D, Shorte SL, Kremer EJ, Brulet P (2005) Visualization of local Ca^{2+} dynamics with genetically encoded bioluminescent reporters. Eur J Neurosci 21(3):597–610. doi:10.1111/j.1460-9568.2005.03871.x
87. Shaner NC, Steinbach PA, Tsien RY (2005) A guide to choosing fluorescent proteins. Nat Methods 2(12):905–909. doi:10.1038/nmeth819
88. Drobac E, Tricoire L, Chaffotte AF, Guiot E, Lambolez B (2010) Calcium imaging in single neurons from brain slices using bioluminescent reporters. J Neurosci Res 88(4):695–711. doi:10.1002/jnr.22249
89. Naumann EA, Kampff AR, Prober DA, Schier AF, Engert F (2010) Monitoring neural activity with bioluminescence during natural behavior. Nat Neurosci 13(4):513–520. doi:10.1038/nn.2518
90. Rodriguez-Garcia A, Rojo-Ruiz J, Navas-Navarro P, Aulestia FJ, Gallego-Sandin S, Garcia-Sancho J, Alonso MT (2014) GAP, an aequorin-based fluorescent indicator for imaging Ca^{2+} in organelles. Proc Natl Acad Sci USA 111(7):2584–2589. doi:10.1073/pnas.1316539111
91. Manjarres IM, Chamero P, Domingo B, Molina F, Llopis J, Alonso MT, Garcia-Sancho J (2008) Red and green aequorins for simultaneous monitoring of Ca^{2+} signals from two different organelles. Pflugers Arch 455(5):961–970. doi:10.1007/s00424-007-0349-5
92. Bakayan A, Vaquero CF, Picazo F, Llopis J (2011) Red fluorescent protein-aequorin fusions as improved bioluminescent Ca^{2+} reporters in single cells and mice. PLoS ONE 6(5):e19520. doi:10.1371/journal.pone.0019520
93. Bakayan A, Domingo B, Miyawaki A, Llopis J (2014) Imaging Ca^{2+} activity in mammalian cells and zebrafish with a novel red-emitting aequorin variant. Pflugers Arch—Eur J Physiol 1–12. doi:10.1007/s00424-014-1639-3
94. Cheung CY, Webb SE, Love DR, Miller AL (2011) Visualization, characterization and modulation of calcium signaling during the development of slow muscle cells in intact zebrafish embryos. Int J Develop Biol 55(2):153–174. doi:10.1387/ijdb.103160cc

95. Martin JR (2012) In vivo functional brain imaging using a genetically encoded Ca^{2+}-sensitive bioluminescence reporter, GFP-Aequorin. In: Martin JR (ed) Genetically Encoded Functional Indicators, vol 72. Neuromethods. Humana Press, New York, pp 1–26
96. Martin JR, Rogers KL, Chagneau C, Brulet P (2007) In vivo bioluminescence imaging of Ca signalling in the brain of Drosophila. PLoS ONE 2(3):e275. doi:10.1371/journal.pone.0000275
97. Rogers KL, Picaud S, Roncali E, Boisgard R, Colasante C, Stinnakre J, Tavitian B, Brulet P (2007) Non-invasive in vivo imaging of calcium signaling in mice. PLoS ONE 2(10):e974. doi:10.1371/journal.pone.0000974
98. Dikici E, Rowe L, Moschou EA, Rothert A, Deo SK, Daunert S (2006) Luminescent proteins: applications in microfluidics and miniaturized analytical systems. In: Photoproteins in bioanalysis. Wiley-VCH Verlag GmbH & Co. KGaA, pp 179–198. doi:10.1002/3527609148.ch10
99. Menon V, Ranganathn A, Jorgensen VH, Sabio M, Christoffersen CT, Uberti MA, Jones KA, Babu PS (2008) Development of an aequorin luminescence calcium assay for high-throughput screening using a plate reader, the LumiLux. Assay Drug Dev Technol 6 (6):787–793. doi:10.1089/adt.2008.0157
100. Tsuji FI, Inouye S, Goto T, Sakaki Y (1986) Site-specific mutagenesis of the calcium-binding photoprotein aequorin. Proc Natl Acad Sci USA 83(21):8107–8111
101. Kurose K, Inouye S, Sakaki Y, Tsuji FI (1989) Bioluminescence of the Ca^{2+}-binding photoprotein aequorin after cysteine modification. Proc Natl Acad Sci USA 86(1):80–84
102. Nomura M, Inouye S, Ohmiya Y, Tsuji FI (1991) A C-terminal proline is required for bioluminescence of the Ca(2+)-binding photoprotein, aequorin. FEBS Lett 295(1–3):63–66
103. Ohmiya Y, Ohashi M, Tsuji FI (1992) Two excited states in aequorin bioluminescence induced by tryptophan modification. FEBS Lett 301(2):197–201
104. Ohmiya Y, Tsuji FI (1993) Bioluminescence of the Ca(2+)-binding photoprotein, aequorin, after histidine modification. FEBS Lett 320(3):267–270
105. Vysotski ES, Lee J (2004) Ca^{2+}-regulated photoproteins: structural insight into the bioluminescence mechanism. Acc Chem Res 37(6):405–415. doi:10.1021/ar0400037
106. Ohmiya Y, Hirano T (1996) Shining the light: the mechanism of the bioluminescence reaction of calcium-binding photoproteins. Chem Biol 3(5):337–347
107. Dikici E, Qu X, Rowe L, Millner L, Logue C, Deo SK, Ensor M, Daunert S (2009) Aequorin variants with improved bioluminescence properties. Protein Eng Des Select (PEDS) 22 (4):243–248. doi:10.1093/protein/gzn083
108. Stepanyuk GA, Golz S, Markova SV, Frank LA, Lee J, Vysotski ES (2005) Interchange of aequorin and obelin bioluminescence color is determined by substitution of one active site residue of each photoprotein. FEBS Lett 579(5):1008–1014. doi:10.1016/j.febslet.2005.01.004
109. Rowe L, Rothert A, Logue C, Ensor CM, Deo SK, Daunert S (2008) Spectral tuning of photoproteins by partnering site-directed mutagenesis strategies with the incorporation of chromophore analogs. Protein Eng Des Select (PEDS) 21(2):73–81. doi:10.1093/protein/gzm073
110. Tsuzuki K, Tricoire L, Courjean O, Gibelin N, Rossier J, Lambolez B (2005) Thermostable mutants of the photoprotein aequorin obtained by in vitro evolution. J Biol Chem 280 (40):34324–34331. doi:10.1074/jbc.M505303200
111. Qu X, Rowe L, Dikici E, Ensor M, Daunert S (2014) Aequorin mutants with increased thermostability. Anal Bioanal Chem 406(23):5639–5643. doi:10.1007/s00216-014-8039-6
112. Rowe L, Ensor M, Mehl R, Daunert S (2010) Modulating the Bioluminescence emission of photoproteins by in vivo site-directed incorporation of non-natural amino acids. ACS Chem Biol 5(5):455–460. doi:10.1021/Cb9002909
113. England PM (2004) Unnatural amino acid mutagenesis: a precise tool for probing protein structure and function. Biochemistry-Us 43(37):11623–11629. doi:10.1021/bi048862q
114. Ryu Y, Schultz PG (2006) Efficient incorporation of unnatural amino acids into proteins in *Escherichia coli*. Nat Methods 3:263–265. doi:10.1038/nmeth864

115. Wang L, Brock A, Herberich B, Schultz PG (2001) Expanding the genetic code of *Escherichia coli*. Science 292(5516):498–500. doi:10.1126/science.1060077
116. Farrell IS, Toroney R, Hazen JL, Mehl RA, Chin JW (2005) Photo-cross-linking interacting proteins with a genetically encoded benzophenone. Nat Methods 2(5):377–384. doi:10.1038/nmeth0505-377
117. Young TS, Ahmad I, Yin JA, Schultz PG (2010) An enhanced system for unnatural amino acid mutagenesis in E. coli. J Mol Biol 395(2):361–374. doi:10.1016/j.jmb.2009.10.030
118. Shrestha S, Paeng IR, Deo SK, Daunert S (2002) Cysteine-free mutant of aequorin as a photolabel in immunoassay development. Bioconjug Chem 13(2):269–275
119. Photoproteins in Bioanalysis (2006) Wiley-VCH, Weinheim
120. Teasley Hamorsky K, Ensor CM, Wei Y, Daunert S (2008) A bioluminescent molecular switch for glucose. Angew Chem Int Ed Engl 47(20):3718–3721. doi:10.1002/anie.200704440
121. Ostermeier M, Nixon AE, Shim JH, Benkovic SJ (1999) Combinatorial protein engineering by incremental truncation. Proc Natl Acad Sci USA 96(7):3562–3567
122. Kanwar M, Wright RC, Date A, Tullman J, Ostermeier M (2013) Protein switch engineering by domain insertion. Methods Enzymol 523:369–388. doi:10.1016/B978-0-12-394292-0.00017-5
123. Stein V, Alexandrov K (2015) Synthetic protein switches: design principles and applications. Trends Biotechnol 33(2):101–110. doi:10.1016/j.tibtech.2014.11.010
124. Scott D, Hamorsky KT, Ensor CM, Anderson KW, Daunert S (2011) Cyclic AMP receptor protein-aequorin molecular switch for cyclic AMP. Bioconjug Chem 22(3):475–481. doi:10.1021/bc100486b
125. Hamorsky KT, Ensor CM, Pasini P, Daunert S (2012) A protein switch sensing system for the quantification of sulfate. Anal Biochem 421(1):172–180. doi:10.1016/j.ab.2011.10.023
126. Gorokhovatsky AY, Rudenko NV, Marchenkov VV, Skosyrev VS, Arzhanov MA, Burkhardt N, Zakharov MV, Semisotnov GV, Vinokurov LM, Alakhov YB (2003) Homogeneous assay for biotin based on *Aequorea victoria* bioluminescence resonance energy transfer system. Anal Biochem 313(1):68–75
127. Rowe L, Deo S, Shofner J, Ensor M, Daunert S (2007) Aequorin-based homogeneous cortisol immunoassay for analysis of saliva samples. Bioconjug Chem 18(6):1772–1777. doi:10.1021/bc070039u
128. Teasley Hamorsky K, Ensor CM, Dikici E, Pasini P, Bachas L, Daunert S (2012) Bioluminescence inhibition assay for the detection of hydroxylated polychlorinated biphenyls. Anal Chem 84(18):7648–7655. doi:10.1021/ac301872u
129. Galvan B, Christopoulos TK (1996) Bioluminescence hybridization assays using recombinant aequorin. Application to the detection of prostate-specific antigen mRNA. Anal Chem 68(20):3545–3550
130. Guenthner PC, Hart CE (1998) Quantitative, competitive PCR assay for HIV-1 using a microplate-based detection system. Biotechniques 24(5):810–816
131. Song X, Coombes BK, Mahony JB (2000) Quantitation of Chlamydia trachomatis 16S rRNA using NASBA amplification and a bioluminescent microtiter plate assay. Comb Chem High Throughput Screening 3(4):303–313
132. Coombes BK, Mahony JB (2000) Nucleic acid sequence based amplification (NASBA) of *Chlamydia pneumoniae* major outer membrane protein (ompA) mRNA with bioluminescent detection. Comb Chem High Throughput Screening 3(4):315–327
133. White SR, Christopoulos TK (1999) Signal amplification system for DNA hybridization assays based on in vitro expression of a DNA label encoding apoaequorin. Nucleic Acids Res 27(19):e25
134. Doleman L, Davies L, Rowe L, Moschou EA, Deo S, Daunert S (2007) Bioluminescence DNA hybridization assay for *Plasmodium falciparum* based on the photoprotein aequorin. Anal Chem 79(11):4149–4153. doi:10.1021/ac0702847

135. Khot PD, Ko DL, Hackman RC, Fredricks DN (2008) Development and optimization of quantitative PCR for the diagnosis of invasive aspergillosis with bronchoalveolar lavage fluid. BMC Infect Dis 8:73. doi:10.1186/1471-2334-8-73
136. Russom A, Ahmadian A, Andersson H, Nilsson P, Stemme G (2003) Single-nucleotide polymorphism analysis by allele-specific extension of fluorescently labeled nucleotides in a microfluidic flow-through device. Electrophoresis 24(1–2):158–161. doi:10.1002/elps.200390008
137. Zerefos PG, Ioannou PC, Traeger-Synodinos J, Dimissianos G, Kanavakis E, Christopoulos TK (2006) Photoprotein aequorin as a novel reporter for SNP genotyping by primer extension-application to the variants of mannose-binding lectin gene. Hum Mutat 27(3):279–285. doi:10.1002/humu.20300
138. Konstantou J, Ioannou PC, Christopoulos TK (2007) Genotyping of single nucleotide polymorphisms by primer extension reaction and a dual-analyte bio/chemiluminometric assay. Anal Bioanal Chem 388(8):1747–1754. doi:10.1007/s00216-007-1383-z
139. Iliadi AC, Ioannou PC, Traeger-Synodinos J, Kanavakis E, Christopoulos TK (2008) High-throughput microtiter well-based bioluminometric genotyping of two single-nucleotide polymorphisms in the toll-like receptor-4 gene. Anal Biochem 376(2):235–241. doi:10.1016/j.ab.2008.02.012
140. Casadei J, Powell MJ, Kenten JH (1990) Expression and secretion of aequorin as a chimeric antibody by means of a mammalian expression vector. Proc Natl Acad Sci USA 87(6):2047–2051
141. Zenno S, Inouye S (1990) Bioluminescent immunoassay using a fusion protein of protein A and the photoprotein aequorin. Biochem Biophys Res Commun 171(1):169–174
142. Erikaku T, Zenno S, Inouye S (1991) Bioluminescent immunoassay using a monomeric Fab'-photoprotein aequorin conjugate. Biochem Biophys Res Commun 174(3):1331–1336
143. Mirasoli M, Michelini, E, DeoS., DikiciE, RodaA, DaunetS(2004) Aequorin fusion proteins as bioluminescent tracers for competitive immunoassays. In: Genetically engineered and optical probes for biomedical applications II, vol 5329. In: Proceedings of SPOE 5329, p 137. doi:10.1117/12.529194
144. Inouye S, Sato J (2008) Comparison of luminescent immunoassays using biotinylated proteins of aequorin, alkaline phosphatase and horseradish peroxidase as reporters. Biosci Biotechnol Biochem 72(12):3310–3313. doi:10.1271/bbb.80524
145. Frank LA (2010) Ca^{2+}-regulated photoproteins: effective immunoassay reporters. Sensors 10 (12):11287–11300. doi:10.3390/s101211287
146. Adamczyk M, Moore JA, Shreder K (2002) Dual analyte detection using tandem flash luminescence. Bioorg Med Chem Lett 12(3):395–398
147. Ito K, Nishimura W, Maeda M, Gomi K, Inouye S, Arakawa H (2007) Highly sensitive and rapid tandem bioluminescent immunoassay using aequorin labeled Fab fragment and biotinylated firefly luciferase. Anal Chim Acta 588(2):245–251. doi:10.1016/j.aca.2007.02.005
148. Desai UA, Wininger JA, Lewis JC, Ramanathan S, Daunert S (2001) Using epitope-aequorin conjugate recognition in immunoassays for complex proteins. Anal Biochem 294(2):132–140. doi:10.1006/abio.2001.5145
149. Qu X, Deo SK, Dikici E, Ensor M, Poon M, Daunert S (2007) Bioluminescence immunoassay for angiotensin II using aequorin as a label. Anal Biochem 371(2):154–161. doi:10.1016/j.ab.2007.08.038
150. Rowe L, Combs K, Deo S, Ensor C, Daunert S, Qu X (2008) Genetically modified semisynthetic bioluminescent photoprotein variants: simultaneous dual-analyte assay in a single well employing time resolution of decay kinetics. Anal Chem 80(22):8470–8476. doi:10.1021/ac801209x
151. Inouye S, Sato J, Sasaki S, Sahara Y (2011) Streptavidin-aequorin fusion protein for bioluminescent immunoassay. Biosci Biotechnol Biochem 75(3):568–571. doi:10.1271/bbb.100798

152. Inouye S, Sato J (2012) Purification of histidine-tagged aequorin with a reactive cysteine residue for chemical conjugations and its application for bioluminescent sandwich immunoassays. Protein Expr Purif 83(2):205–210. doi:10.1016/j.pep.2012.04.001
153. Daunert S, Barrett G, Feliciano JS, Shetty RS, Shrestha S, Smith-Spencer W (2000) Genetically engineered whole-cell sensing systems: coupling biological recognition with reporter genes. Chem Rev 100(7):2705–2738
154. Feliciano J, Pasini P, Deo SK, Daunert S (2008) Photoproteins as reporters in whole-cell sensing. In: Deo SK (ed) Protein science encyclopedia. Wiley-VCH, New York, pp 131–154
155. Kozlova O, Zwinderman M, Christofi N (2005) A new short-term toxicity assay using *Aspergillus awamori* with recombinant aequorin gene. BMC Microbiol 5:40. doi:10.1186/1471-2180-5-40
156. Zeinoddini M, Khajeh K, Behzadian F, Hosseinkhani S, Saeedinia AR, Barjesteh H (2010) Design and characterization of an aequorin-based bacterial biosensor for detection of toluene and related compounds. Photochem Photobiol 86(5):1071–1075. doi:10.1111/j.1751-1097.2010.00775.x
157. Rider TH, Petrovick MS, Nargi FE, Harper JD, Schwoebel ED, Mathews RH, Blanchard DJ, Bortolin LT, Young AM, Chen J, Hollis MA (2003) A B cell-based sensor for rapid identification of pathogens. Science 301(5630):213–215. doi:10.1126/science.1084920
158. Araki N, Iida M, Amino N, Morita S, Ide A, Nishihara E, Ito M, Saito J, Nishikawa T, Katsuragi K, Miyauchi A (2015) Rapid bioassay for detection of thyroid-stimulating antibodies using cyclic adenosine monophosphate-gated calcium channel and aequorin. EurThyroid J 4(1):14–19. doi:10.1159/000371740
159. Date A, Pasini P, Daunert S (2010) Integration of spore-based genetically engineered whole-cell sensing systems into portable centrifugal microfluidic platforms. Anal Bioanal Chem 398(1):349–356. doi:10.1007/s00216-010-3930-2
160. Bjerketorp J, Hakansson S, Belkin S, Jansson JK (2006) Advances in preservation methods: keeping biosensor microorganisms alive and active. Curr Opin Biotechnol 17(1):43–49. doi:10.1016/j.copbio.2005.12.005
161. Date A, Pasini P, Daunert S (2007) Construction of spores for portable bacterial whole-cell biosensing systems. Anal Chem 79(24):9391–9397. doi:10.1021/ac701606g
162. Date A, Pasini P, Daunert S (2010) Fluorescent and bioluminescent cell-based sensors: strategies for their preservation. Adv Biochem Eng Biotechnol 117:57–75. doi:10.1007/10_2009_22
163. Date A, Pasini P, Sangal A, Daunert S (2010) Packaging sensing cells in spores for long-term preservation of sensors: a tool for biomedical and environmental analysis. Anal Chem 82(14):6098–6103. doi:10.1021/ac1007865
164. Weissleder R (2001) A clearer vision for in vivo imaging. Nat Biotechnol 19(4):316–317. doi:10.1038/86684
165. Frangioni JV (2003) In vivo near-infrared fluorescence imaging. Curr Opin Chem Biol 7(5):626–634
166. Gao X, Cui Y, Levenson RM, Chung LW, Nie S (2004) In vivo cancer targeting and imaging with semiconductor quantum dots. Nat Biotechnol 22(8):969–976. doi:10.1038/nbt994
167. Lam CW, James JT, McCluskey R, Arepalli S, Hunter RL (2006) A review of carbon nanotube toxicity and assessment of potential occupational and environmental health risks. Crit Rev Toxicol 36(3):189–217
168. Yu WW, Chang E, Drezek R, Colvin VL (2006) Water-soluble quantum dots for biomedical applications. Biochem Biophys Res Commun 348(3):781–786. doi:10.1016/j.bbrc.2006.07.160
169. Liu Z, Davis C, Cai W, He L, Chen X, Dai H (2008) Circulation and long-term fate of functionalized, biocompatible single-walled carbon nanotubes in mice probed by Raman spectroscopy. Proc Natl Acad Sci USA 105(5):1410–1415. doi:10.1073/pnas.0707654105
170. Bakayan A, Domingo B, Miyawaki A, Llopis J (2014) Imaging Ca activity in mammalian cells and zebrafish with a novel red-emitting aequorin variant. Eur J Physiol, Pflugers Archiv. doi:10.1007/s00424-014-1639-3

171. Grinstead KM (2015) Aequorin mutants with site-specifically incorporated non-natural amino acids for biomedical applications. Dissertation, University of Miami, Open Access Dissertations
172. Mekahli D, Bultynck G, Parys JB, De Smedt H, Missiaen L (2011) Endoplasmic-reticulum calcium depletion and disease. Cold Spring Harbor Perspect Biol 3(6). doi:10.1101/cshperspect.a004317
173. Sama DM, Norris CM (2013) Calcium dysregulation and neuroinflammation: discrete and integrated mechanisms for age-related synaptic dysfunction. Ageing Res Rev 12(4):982–995. doi:10.1016/j.arr.2013.05.008

Adv Biochem Eng Biotechnol (2016) 154: 181–200
DOI 10.1007/10_2015_337

Published Online: 17 October 2015

Whole-Cell Biosensors as Tools for the Detection of Quorum-Sensing Molecules: Uses in Diagnostics and the Investigation of the Quorum-Sensing Mechanism

Gregory O'Connor, Leslie D. Knecht, Nelson Salgado, Sebastian Strobel, Patrizia Pasini and Sylvia Daunert

Abstract Genetically engineered bacterial whole-cell biosensors are powerful tools that take advantage of bacterial proteins and pathways to allow for detection of a specific analyte. These biosensors have been employed for a broad range of applications, including the detection of bacterial quorum-sensing molecules (QSMs). Bacterial QSMs are the small molecules bacteria use for population density-dependent communication, a process referred to as quorum sensing (QS). Various research groups have investigated the presence of QSMs, including *N*-acyl homoserine lactones (AHLs) and autoinducer-2 (AI-2), in physiological samples in attempts to enhance our knowledge of the role of bacteria and QS in disease states. Continued studies in these fields may allow for improved patient care and therapeutics based upon QSMs. Furthermore, bacterial whole-cell biosensors have elucidated the roles of some antibiotics as QS agonists and antagonists.

G. O'Connor · L.D. Knecht (✉) · N. Salgado · S. Strobel · P. Pasini · S. Daunert
Department of Biochemistry and Molecular Biology, University of Miami,
Miller School of Medicine, Miami, FL 33136, USA
e-mail: lknecht@miami.edu

L.D. Knecht
Department of Chemistry, University of Miami, Miami, FL 33146, USA

Graphical Abstract

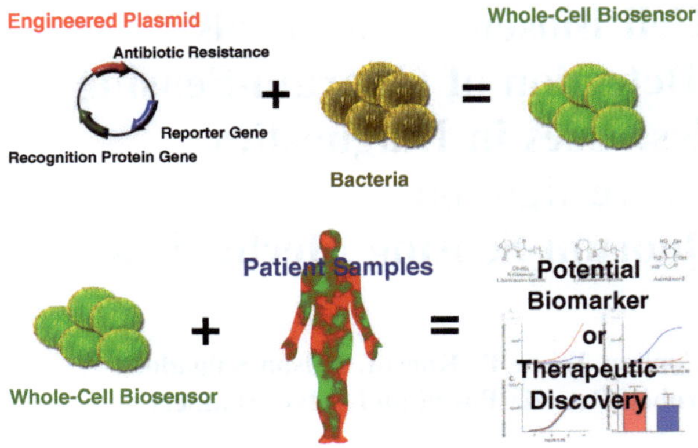

Keywords Autoinducer-2 (AI-2) · Cystic Fibrosis (CF) · Inflammatory Bowel Disease (IBD) · *N*-acyl homoserine lactones (AHLs) · Quorum-sensing (QS) · Quorum-sensing molecules (QSMs) · Urinary Tract Infections (UTI) · Whole-cell biosensors

Contents

1 Introduction

Quorum sensing (QS) is the population-dependent exchange of small chemical signaling molecules that allow bacteria to monitor their population density and regulate gene expression accordingly. The phenomenon was first described in the early 1970s with the microorganism *Vibrio harveyi*, which employs QS to control the emission of light (bioluminescence) [1, 2]. Traditionally, these signaling molecules have been classified into two subgroups: *N*-acyl homoserine lactones (AHLs, Fig. 1), which are used by Gram-negative bacteria, and oligopeptides, which are used by Gram-positive bacteria. In recent years, however, it has been found that quorum-sensing molecules (QSMs) are not limited to only these two subgroups, with the discovery of pseudomonas quinolone signal (PQS), autoinducer 2 (AI-2, Fig. 1), and autoinducer 3 (AI-3), among others; interestingly, some of these molecules, such as AI-2, can appear in both Gram-positive and Gram-negative bacteria [3, 4]. QSMs are estimated to influence 4–10 % of the bacterial genome and 20 % of the proteome, regulating myriad genes ranging from factors responsible for bacterial virulence to genes involved in basic metabolic processes [5]. For example, QSMs allow bacteria to communicate with each other to form biofilms, an extracellular matrix system that helps protect the bacteria: most bacteria reside in these three-dimensional matrix systems rather than existing in the planktonic form [6]. In order for bacteria to be able to respond to QSMs, quorum-sensing systems comprise recognition/regulatory proteins that are able to bind to the QSMs and regulate gene expression, as well as synthases which allow for production of QSMs. Table 1 provides examples of quorum-sensing organisms, their regulatory proteins, and QSMs.

Previously, studies on QS and the mechanism of action of the QSMs mostly relied on the native bacteria and examining growth rates and up- or down-regulation of specific genes. Although these methods would give qualitative information about the effects of QSMs on the microorganisms, the amount of quantitative information was limited. With the identification of QS regulatory proteins and their promoter binding regions, genetically engineered whole-cell bacterial biosensors (also called

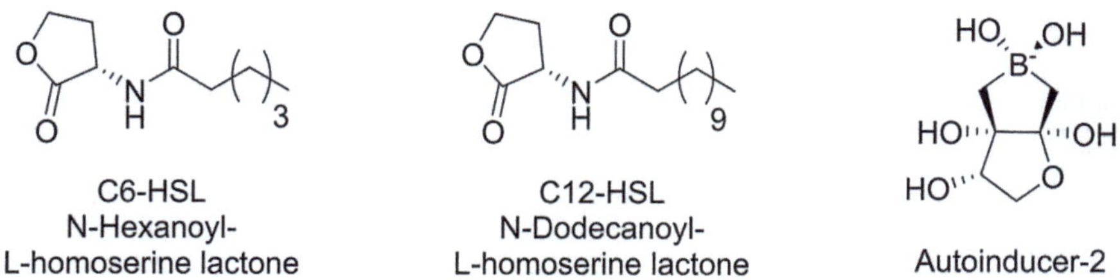

Fig. 1 Chemical structures of *N*-hexanoyl-*L*-homoserine lactone, *N*-dodecanoyl-*L*-homoserine lactone, and autoinducer-2 (boron form) QSMs

Table 1 Organisms and their QSMs that have been utilized in whole-cell biosensors

Organism	QSM	Synthase	Receptor/regulator
Pseudomonas aeruginosa	C4-HSL	RhlI	RhlR
	3-OXO-C12-HSL	LasI	LasR, QscR
	HHQ, PQS	PqsA	PqsR
Vibrio fischeri	3-oxo-C6-HSL	LuxI	LuxR
	AI-2	AinS	AinR
	CAI-1	LuxS	LuxP
Agrobacterium tumefaciens	3-oxo-C8-HSL	TraI	TraR

biosensing systems) have emerged as tools for detection of QSMs and understanding of the quorum-sensing mechanism [7]. These biosensing systems generate a detectable response under certain conditions, often times dose-dependent upon recognition of a specific analyte of interest. Whole-cell bacterial biosensors employ bacteria that have been genetically modified to contain a biological recognition agent coupled with a reporter. This genetic modification generally occurs in two ways: insertion of a genetically engineered plasmid and/or modification of the bacterial genome.

Developing a plasmid for use in a biosensing system allows for customization of a sensor depending upon the need, thus there is a large degree of variability. Several key elements that are required and/or commonly employed include genes for: a reporter protein, promoter region, antibiotic resistance, and over expression of the regulatory protein. The reporter plays a large role in the sensitivity of the biosensor, because it is the recognition agent for the assay. The reporter protein(s) selected for bacterial whole-cell biosensors are generally ones that can be detected optically, via an instrument or the naked eye (bioluminescence, chemiluminescence, fluorescence, and colorimetry), or electrochemically. Table 2 contains a list of a few common reporters, along with their detection methods and some advantages and disadvantages. The promoter region determines the production of the reporter protein, as well as playing a role in the specificity and selectivity of the system, thus it should be selected based upon the analyte of interest for the sensor. Including antibiotic resistance under a constitutive promoter allows for easy selection of bacteria expressing the plasmid, thus preventing the loss of plasmid as the bacterial culture grows. Overexpression of a regulatory protein increases the sensor's ability

Table 2 Commonly used reporter proteins for bacterial whole-cell-based sensing systems

Gene	Protein	Detection Method	Advantages	Disadvantages
lux	Bacterial luciferase	Bioluminescence	Ease of measurement, rapid response	Heat lability, oxygen requirement
luc	Firefly luciferase	Bioluminescence	High sensitivity, no heat lability	Exogenous substrate requirement, oxygen requirement
gfp	Green fluorescent protein (jellyfish)	Fluorescence	No substrate requirement	Low sensitivity, broad absorption spectra
lacZ	β-Galactosidase (*E. coli*)	Chemiluminescence, bioluminescence, fluorescence, colorimetry, electrochemistry	Wide variety of detection methods, detection by naked eye	Exogenous substrate requirement

Adapted from Raut et al. [7]

to detect the analyte of interest: the more protein you have, the higher its chances of encountering the analyte. Once a plasmid has been genetically engineered, it is a relatively simple task to insert the DNA into electro- or chemically competent bacterial cells, thus creating a whole-cell bacterial biosensor.

In addition to genetically engineering a plasmid to employ in a whole-cell biosensor, it is possible to modify the bacterial genome to exploit the original organism itself as a whole-cell biosensor. Many bacterial strains generate QSMs through various synthesis pathways, thus modifications can be made to bacterial genomes to prevent the generation of and/or the response to QSMs. One such example is the *V. harveyi* strain BB170; native *V. harveyi* activates its bioluminescence in response to AHLs and AI-2 and can produce these QSMs. The BB170 strain was genetically modified to have its AHL detection pathway, as well as its ability to synthesize AHLs and AI-2, blocked. Thus BB170 was successfully engineered to only generate bioluminescence in response to exogenous AI-2, allowing it to function as a biosensing system.

Regardless of the method used to develop the sensing system, bacterial whole-cell biosensors are categorized into two general types based upon whether expression of the reporter protein is inducible or constitutive. Inducible expression systems employ cells that have been genetically modified to contain an inducible promoter fused to a reporter gene. In the absence of the analyte (inducer), the reporter gene is expressed at low basal levels, determining the "blank" level of the system. When the analyte (inducer) is present, the reporter gene is expressed in a dose-dependent manner within a certain range of concentrations, allowing for a standard curve to serve as a control for sample measurements. Reporter gene expression can be negatively or positively regulated, defined by how the analyte–protein complex interacts with the gene promoter. In negative regulation (Fig. 2a), the regulatory protein is bound to the operator/promoter region, thus preventing reporter and other downstream gene expression. The analyte can bind the regulatory protein, causing the complex to be released from the promoter region and allowing for gene expression to occur. In positive regulation (Fig. 2b), the unbound regulatory protein complexes with the analyte and this complex can then bind to the operator/promoter region to activate reporter gene expression.

In contrast to inducible expression, constitutive expression systems are based upon generation of high basal levels of reporter expression (Fig. 2c). This approach is primarily employed in bacterial whole-cell biosensors as a measure of toxicity: however, nonspecific increased toxicity leads to cell death, reducing the amount of reporter protein and thus the overall signal. Therefore overall toxicity can be recorded, although it does not provide specific information about the nature and mode of action of the toxic agent(s) that is (are) killing the cells.

Bacterial whole-cell biosensors possess several advantages that have increased their popularity in recent years: they are relatively easy and inexpensive to prepare and store, are robust with a tendency to be stable to environmental changes such as pH or temperature fluctuations, can be engineered for multiplexing by using different recognition–reporter pairs, and can be integrated into various platforms allowing for high-throughput or on-site detection. In addition, regulatory proteins

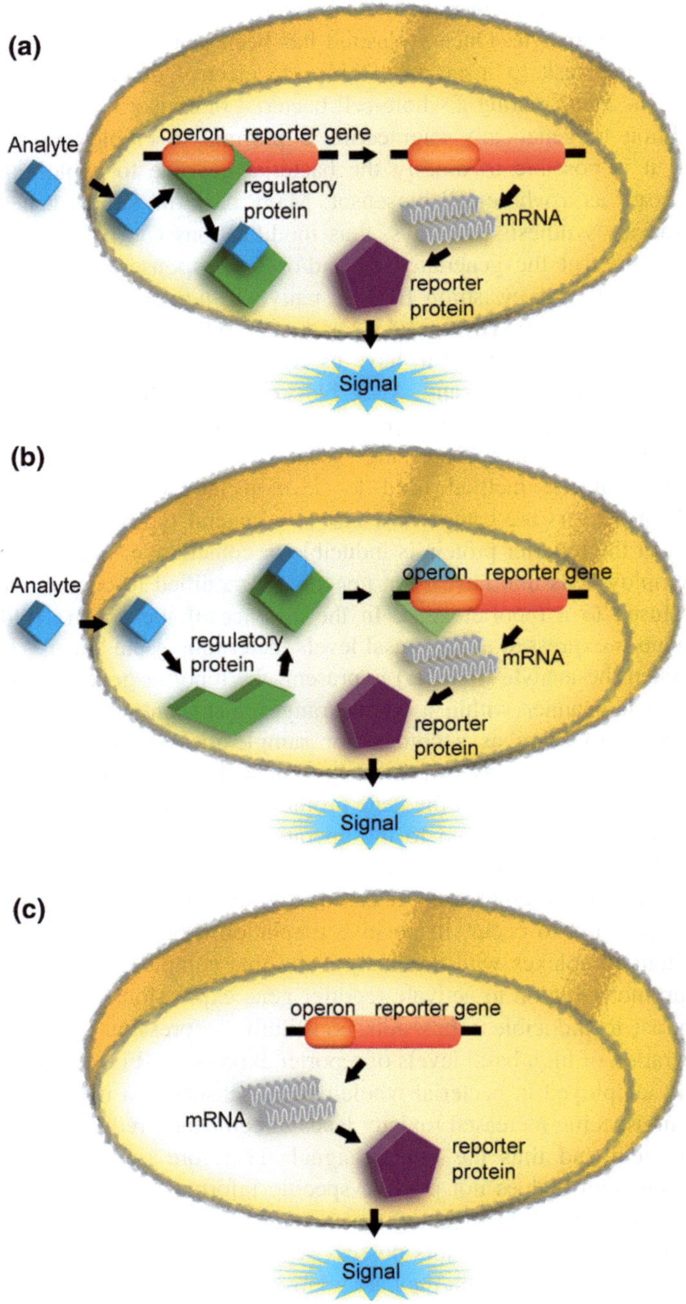
(a)
Analyte
operon
reporter gene
regulatory protein
mRNA
reporter protein
Signal
(b)
Analyte
operon
reporter gene
regulatory protein
mRNA
reporter protein
Signal
(c)
operon
reporter gene
mRNA
reporter protein
Signal

◀ **Fig. 2** Whole-cell-based biosensing system types. **a** In positive regulation of reporter gene expression, the regulatory protein is bound to the operon, preventing expression of the reporter gene. The analyte enters the cell and binds the regulatory protein, allowing for reporter gene expression. **b** In negative regulation of reporter gene expression, the analyte enters the cell and binds the regulatory protein. This analyte-regulatory protein complex can bind the operon and allow for expression of the reporter gene. **c** In constitutive expression, the reporter gene is constantly expressed

may prove difficult to express and purify, thus whole-cell biosensors provide an alternative route to investigating interactions with these proteins. Interestingly, bacterial whole-cell-based biosensors can provide physiologically relevant data and evaluate the bioavailability of the analyte of interest, inasmuch as the target chemical must be ingested by the cells in order to trigger a response. However, this requirement can also prove to be a limitation, as the bacteria rely primarily on diffusion, and thus may lack specific transport channels that may be necessary for the uptake of certain analytes.

2 QSMs in Physiological Samples via Bacterial Whole-Cell-Based Biosensing Systems

2.1 AHL Detection

Studies have shown that bacterial AHLs, such as 3-oxo-dodecanoyl-homserine lactone (3-oxo-C12-HSL), not only regulate bacterial virulence but can also stimulate cells important for inflammation and immune defense in the host [8]. It has been demonstrated that in burn wound mouse models infected with QS deficient bacterial strains, there was no production of mRNA for pro-inflammatory cytokines IL-6, TNFα, CGB, and TGFβ [9]. The opportunistic pathogen *Pseudomonas aeruginosa*, which employs QS to regulate virulence factor production and biofilm formation, plays an important role in chronic wounds, which are a large burden to the health system worldwide; *P. aeruginosa* is estimated to be present in up to 80 % of chronic wound cases. In addition, when *P. aeruginosa* is present, wounds are more severe and heal at a slower rate, which is thought to be caused by cell-to-cell communication with gram-positive bacteria, including *Staphylococcus aureus* [10–12].

The implications of bacteria in a variety of disorders have prompted researchers to investigate the role of QS in certain disease states by detecting QSMs in physiological samples. In an effort to elucidate the roles of QSMs in their natural environments, physico-chemical methods have been devised for their detection, including but not limited to high-performance liquid chromatography (HPLC), reversed phase HPLC mass spectrometry (RP-HPLC-MS), ultra-high-resolution mass spectrometry, and gas chromatography mass spectrometry (GC-MS) [13–15]. Additionally, work by Campagna et al. [16] and May et al. [17] has led to liquid chromatography–tandem mass spectrometry (LC-MS/MS) detection methods for

Table 3 An overview of bacterial whole-cell-based biosensors employed for detection of AHLs in physiological samples

Plasmid(s)	Sensor name	Bacterial strain	Detects	Promoter	Reporter
pSB401		*E. Coli*	Short-chain AHLs	*luxRI*	LuxCDABE
pSB1075		*E. Coli*	Long-chain AHLs	*lasRI*	LuxCDABE
pECP61.5		*P. aeruginosa*	Short-chain AHLs	*rhlA*	LacZ
pKDT17		*P. aeruginosa*	Long-chain AHLs	*lasB*	LacZ
pCF218, pMV26	A136	*A. tumefaciens*	Short- and long-chain AHLs	*traRG*	LacZ
pSB406		*E. Coli*	Short chain-AHLs	*rhlRI*	LuxCDABE
	CV026	*C. violaceum*	Short chain-AHLs	*luxI*-box-like	Violacein
	VIR07	*C. violaceum*	Long chain-AHLs	*cviR*	Violacein

AI-2 and AHLs in biologically relevant samples. Unfortunately, these methods require extensive sample extraction and/or preparation, large amounts of trained technician time, and expensive instrumentation, thus limiting their utility. Therefore we explore some of the findings from utilizing bacterial whole-cell-based biosensors for QSMs in physiological samples. The sensors employed for AHL detection are listed in Table 3. Please note, for the experiments and sensors described in the Table 3, short chain AHLs refer to AHLs with carbon chains of seven or less carbons, whereas long-chain AHLs refer to AHLs with carbon chains of eight or more carbons.

2.1.1 Cystic Fibrosis (CF)

Cystic fibrosis (CF) is an autosomal recessive genetic disorder that primarily affects the lungs, as well as the pancreas, liver, and intestine. Common symptoms include infertility, poor growth, sinus infections, and lung infections that lead to difficulty breathing. Although a large number of bacteria have been indicated to play a role in CF, the pathogenesis has yet to be fully elucidated [18]. Studies have shown that *P. aeruginosa* forms thick biofilms within the lungs of CF patients and uses QS extensively to modulate both biofilm formation and maturation [19, 20]. Of these bacteria, *P. aeruginosa* and *Burkholderia cepacia*, which are known to colonize the lungs of CF patients, both use AHL-dependent QS regulation [21].

In a study by Middleton et al. [22], sputum samples were collected from stable CF patients, extracted, and then analyzed by employing two *E. coli* whole-cell-based sensors bearing plasmids pSB401 and pSB1075, to detect short- and long-chain AHLs, respectively. For sputum samples colonized with *P. aeruginosa*, 71 % contained short-chain AHLs and 61 % contained long-chain AHLs ($n = 42$). Correspondingly, for sputum samples colonized with *B. cepacia*, 87 % contained short-chain AHLs and 50 % contained long-chain AHLs ($n = 8$). Another study published the same year by Erickson et al. [23] further confirmed the presence of AHLs in sputum samples from CF patients. Whole-cell sensors based on *P. aeruginosa* carrying plasmids pECP61.5 and pKDT17 were utilized to detect short- and long-chain AHLs, respectively. Short-chain AHLs were found in 26 % of the patient sputum samples and long-chain AHLs were found in 78 % of the samples ($n = 29$). This finding is in contrast to the results of Middleton et al. [22], however there are several possible explanations: small numbers of different patients were assessed, different sample processing/extraction methods were used, and different sensing systems, both the bacterial strains and the plasmids employed, were used. Unfortunately, the small number of patients assayed, the variability of the patient disease status, and the lack of an established range of AHLs in sputum, do not allow for conclusions to be drawn on how and if AHL levels fluctuate in CF patient samples.

These works were followed by Chambers et al. [24], which focused on identifying a broad range of AHLs within mucopurulent respiratory secretions from CF patients undergoing lung transplantation. *Agrobacterium tumefaciens*-based whole-cell sensor A136 was employed, which contains dual plasmids pCF218 and pMV26, allowing for detection of both short- and long-chain AHLs. Mucopurulent material was extracted, separated by reversed-phase fast pressure liquid chromatography (FPLC), and then analyzed by sensor A136. By combining the whole-cell-based sensor with FPLC, AHLs were found in 69 % of patient samples, with seven unique AHLs being identified. These studies have helped to establish bacterial whole-cell-based biosensors as important tools for QSM detection, as well as have provided researchers a focal point for further discoveries to improve the lives and care of patients with CF.

2.1.2 Inflammatory Bowel Disease (IBD)

IBD is an autoimmune disorder that is characterized by a group of chronic inflammatory conditions of the gastrointestinal tract (GI), commonly causing severe abdominal pain, vomiting, diarrhea, rectal bleeding, and weight loss. IBD can be further classified based upon the symptoms and the extent of the inflammation within the GI tract, with Crohn's and Ulcerative Colitis being the two principal types. IBD is thought to be caused in genetically predisposed individuals by an overly aggressive immune response to commensal bacteria [25], thus research in this field has focused on the gut microbiome. The human gut is host to a broad range of bacteria, and although QS has been demonstrated for a limited number of intestinal bacteria, mainly pathogens, it is likely that a wide variety of enteric microorganisms, including commensal bacteria, communicate with each other and the host by means of QSMs.

Studies have shown that the bacteria of the gut play a role in IBD etiology, thus investigating QSMs in individuals may grant a greater understanding of the disease. To this end, work by Kumari et al. [15, 26] has led to the development and optimization of *E. coli* based whole-cell sensing systems, bearing plasmids pSB406 and pSB1075 [27], for quantitative detection of short- and long-chain AHLs (Fig. 3a, b), respectively, in human samples such as saliva and stool. By focusing on saliva and stool samples, this allows for a detection method that is completely non-invasive to patients. It is important to note that these methods do not require sample extraction or extensive sample preparation; simple centrifugation or dilution in water is all that is necessary to prepare samples. Additionally, these optimized systems allow for nanomolar limits of detection in physiological matrices, which makes these systems relevant, due to nanomolar levels of QSMs being necessary to initiate cellular communication [7]. Saliva from IBD and healthy individuals were assayed, revealing greatly varying levels of AHLs across individuals that were consistent for each individual for a period of time. Furthermore, stool samples from newborns admitted to a Neonatal Intensive Care Unit (NICU) also contained varying levels of AHLs. These results revealed for the first time that QSMs can be detected in human saliva and stool. These findings are of added interest, because it is well known that QSMs such as AHLs are subject to degradation by environmental factors, such as pH and temperature, and may be degraded by enzymes

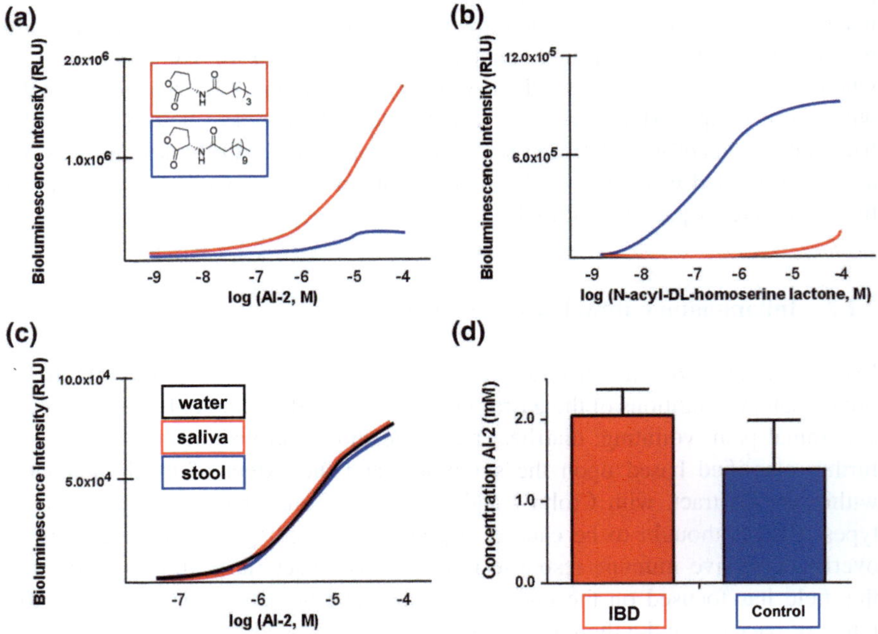

Fig. 3 Dose response curves generated for *N*-hexanoyl-DL-homoserine lactone (*red*) and *N*-dodecanoyl-DL-homoserine lactone (*blue*) on *E. coli* whole-cell-based biosensor harboring plasmid **a** pSB406 or **b** pSB1075, indicating the selectivity of each sensor for short or long chain AHLs (adapted from Kumari et al. [26]). **c** Dose response curves of AI-2 generated in water (*black*), 1:100 (v/v) diluted saliva (*red*), and 1:750 (w/v) stool (*blue*) with the *V. harveyi* BB170 sensor, indicating optimized sample dilutions for use with the sensor. **d** Saliva from IBD patients (*red*) and healthy controls (*blue*) were then diluted and assayed with the sensor, $n = 3$ (adapted from Raut et al. [30])

present in the gut [28, 29]. If a correlation can be established between the health of individuals and the levels of QSMs found in their samples, QSMs may develop into a biomarker for IBD, as well as other bacterial disorders. Current studies in the Daunert laboratory are in progress in which physiological samples from IBD patients, as well as matched controls, are being analyzed for their QSM levels, in order to investigate such correlation with disease status.

2.1.3 Hospital-Acquired Urinary Tract Infections (UTI)

Nosocomial infections (NI), commonly known as hospital-acquired infections, include multiple types of fungal and bacterial infections that are aggravated by the weakened resistance of patients. With *P. aeruginosa* known as a common link among many types of NI, multiple groups have worked to characterize the AHL profile of this microorganism during infection [31–33]. Gupta et al. examined the role of QSMs for their ability to induce pro- and anti-inflammatory renal cytokines and inflammation in an experimental urinary tract infection (UTI) model [34]. Their results suggested that deficiency for production of QSMs in mutant strains of *P. aeruginosa* resulted in non-stimulation of a pro-inflammatory response. Kumar et al. [31] employed sensors pKDT17, A136, CV026, and pECP61.5 to detect a broad range of AHLs: long-chain, both long- and short-chain, short-chain, and short-chain, respectively. Analysis of *P. aeruginosa* isolates from catheter-associated UTI found AHL molecules of varying types in all samples, indicating that the virulence factors activated may vary between UTIs. These findings were verified in work by Lakshmana Gowda et al. [33], who applied sensors pKDT17 and A136, not only to UTI isolates, but to other nosocomial infection *P. aeruginosa* isolates as well. However, work by Senturk et al. [32], which utilized sensors CV026 and VIR07 for detection of short- and long-chain AHLs, respectively, revealed another side to *P. aeruginosa* infection; although the diversity of AHLs seen previously was detected in most UTI isolates, 7 % contained no detectable levels of AHLs. Sequence analysis revealed mutations within genes crucial to the AHL pathway; *lasI*, *lasR*, *rhlR*, and *rhlI*. The authors hypothesized that these findings signify that *P. aeruginosa* possess additional virulence pathways that are independent of AHLs. Continued analysis of QSM levels in UTI isolates may help to elucidate the mechanism of *P. aeruginosa* infection, allowing for greater patient care and treatment.

2.1.4 Paper Strip Detection of Quorum-Sensing Molecules

The previously mentioned research on AHL detection has been focused on assays performed in the laboratory, however, when developing sensors that may have relevance for diagnostics or environmental detection, it is important to utilize a type of sensor and detection method which can be portable, that is, applied in a bedside or field kit. Whole-cell biosensors are ideal candidates due to their ability to withstand a range of environmental conditions, requirement of minimal or no sample pretreatment, easy and rapid detection, and cost-effectiveness, to name a few [35, 36]. Due to the prevalence of QS in the pathogenesis of bacteria-related disorders, developing a fast, reliable, visual detection method for bacterial signaling molecules is of great interest. Work has been done to sense AHLs visually without instrumentation; however, these methods have shortcomings such as being time-consuming and not allowing detection in a dose-dependent manner [37]. To overcome these shortcomings, Struss et al. [38] designed a whole-cell biosensor that was amenable to paper strip incorporation for the semi-quantitative detection of bacterial quorum-signaling molecules. This whole-cell-based biosensor employed *E. coli* cells harbouring plasmid pSD908, which detects long-chain AHLs and contains the reporter gene *lacZ*, encoding β-galactosidase. β-galactosidase activity can be determined using a variety of substrates and, depending on the chosen substrate, detection of the analytical signal can be fluorescent, chemiluminescent, electrochemical, or colorimetric [26]. The *E. coli* cells harboring plasmid pSD908 were immobilized on the filter-paper-based strips by suspending the cells in a drying protectant solution and applying the suspension directly to filter-paper strips to allow them to dry via a liquid-drying method. This approach was used to help in the long-term preservation of cells. After the cells were dried on the filter-paper strips, they could then be immersed in liquid samples for detection of AHLs. After the sample immersion step, the strips were incubated in an 5-bromo-4-chloro-3-indolyl-β-D-galactoside (X-gal) solution to develop a blue color. The intensity of the blue color increased with an increasing concentration of AHL, thus allowing semi-quantitative naked-eye evaluation (Fig. 4a). The paper strip sensor was employed for detection of AHLs in saliva and AHL levels were found in both healthy and diseased individuals (Fig. 4b), mirroring previous findings and proving the paper strip sensor's ability to detect AHLs in physiological samples. It is

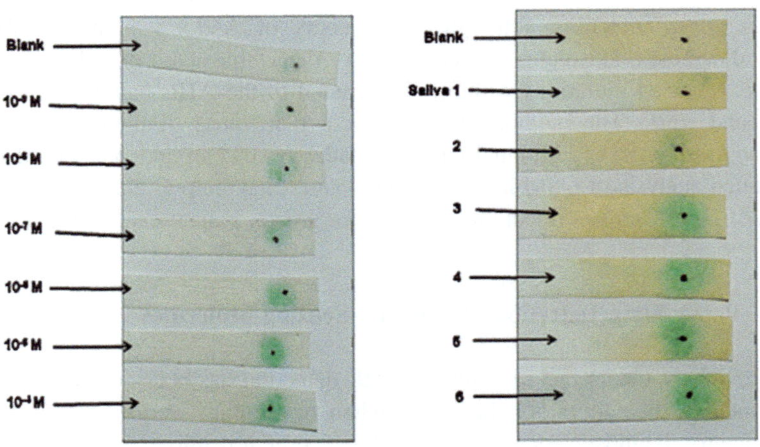

Fig. 4 **a** Dose-response data in buffer using filter-paper-based strips. The color intensity increases with increasing concentration of AHL. **b** Detection of AHLs in healthy patients (*1–5*) and a patient with Crohn's Disease (*6*). Reproduced with copyright permission from Struss et al. [38]

important to note that the assay was also performed at room temperature (RT), which demonstrates the applicability of the system for point-of-care diagnostics and field applications. Additionally, the paper strip sensors were found to be stable for at least three months at 4 °C without loss of analytical characteristics. Thus continued work with this and other bacterial whole-cell-based biosensors may allow for commercially available point-of-care and field tests for QSMs.

2.2 *AI-2 Detection*

In recent years, a group of molecules referred to as autoinducer-2 (AI-2) have been found in both Gram-positive and Gram-negative bacteria [39]. This finding is of great importance, as this indicates AI-2 molecules are used for interspecies communication. The production of these molecules from their common precursor, 4,5-dihydroxy-2,3-pentanedione (DPD), is catalyzed by the LuxS enzyme. In the organism *V. harveyi*, the AI-2 diffuses out of and into the cell and binds to a periplasmic binding protein, LuxP. This binding event starts a cascade of phosphorylation/dephosphorylation events and leads to the expression of luciferase for the production of light [40].

2.2.1 Inflammatory Bowel Disease (IBD)

As mentioned previously, AHLs are currently being investigated as a potential biomarker for IBD. As part of this research, Kumari et al. [26] and Raut et al. [30] hypothesized that monitoring the QSMs in physiological samples over time can give an idea of the amount of bacteria present, inflammation, and the progress of the disease. Specifically, as part of determining QSM profiles in patients, the interspecies communication molecule AI-2, which encompasses both Gram-negative and Gram-positive bacteria, could prove an ideal non-invasive biomarker for gastrointestinal disease.

Vibrio harveyi strain BB170 has been genetically modified to respond only to AI-2. Previously, this strain had only been used as a bioassay to screen cell cultures and identify bacteria able to produce AI-2. Raut et al. [30] developed and optimized a biosensing system based on *V. harveyi* BB170 and applied it for the quantitative detection of AI-2 in saliva, stool, and intestinal samples from IBD patients (Fig. 3c, d). As with their previous AHL research, a focus was placed on developing a non-invasive assay that did not require extensive sample preparation or extraction. With a similar process of simple centrifugation and dilution of samples in water, stool and saliva samples could be assayed for AI-2. Stool samples from IBD patients were then analyzed and found to exhibit considerably varied levels of AI-2. The authors hypothesized that these variations may reflect perturbations in the microflora of the inflamed intestines [41, 42]. AI-2 was also detectable in saliva samples from IBD patients via the BB170 biosensor in nanomolar concentrations, which correlates well with the levels of AI-2 previously detected in human saliva via LC-MS/MS (244-965 nM) [16]. Thus continued work with the universal signaling molecule AI-2 may help us better understand and diagnose IBD.

3 Quorum-Sensing Agonists and Antagonists

As our knowledge of QSMs and their potential roles as biomarkers of disease states has expanded, interest has focused on molecules that interfere with the quorum-sensing network, either by acting as agonists or antagonists [43]. By blocking or mimicking the native quorum-sensing molecule's signal, more insight can be gained in the quorum-sensing molecular mechanism and it may be possible to interfere effectively with QS and modify the behavior of bacteria. These studies typically demonstrated that compounds analogous to the cognate quorum-sensing molecule (i.e. having a lactone ring but varying tail lengths, oxidation states, or saturation levels) induced weak to moderate gene expression [44]. Whole-cell biosensors have proven to be valuable tools to elucidate effects of structurally similar and dissimilar molecules on QS.

3.1 Antibiotics

Antibiotics are employed all over the world for the treatment and prevention of bacterial infection, however, their mechanisms of action are not fully elucidated. To this end, researchers have investigated the effects of antibiotics on QS pathways. Previous reports demonstrate that the antibiotic azithromycin was able to interfere with the regulation of the quorum-sensing network of *P. aeruginosa,* inhibiting virulence factors and biofilm formation [45]. These studies, however, only investigated the behavior changes of the native cells, such as motility and biofilm formation, but did little to investigate the mechanism of action of the antibiotic on the quorum-sensing system. To better understand how antibiotics are able to interact with the quorum-sensing network and screen for antibiotics that are able to interfere with QS, Struss et al. employed *E. coli* whole-cell biosensors harboring plasmids pSB406 and pSB1075 for the detection of short- and long-chain AHLs, respectively [26, 46]. The authors investigated three antibiotics that were commonly used in the treatment of IBD: ciprofloxacin, metronidazole, and tinidazole. Azithromycin was used as a control, in addition to exogenous AHL addition, as it has previously been shown to act as a quorum-sensing agonist/antagonist [47–49].

As expected, azithromycin was able to induce a dose-dependent response in both the pSB406- and pSB1075-based whole-cell sensors. Furthermore, this response occurred in both the presence and absence of exogenous cognate AHL. The antibiotics tested were also able to activate the systems in both whole-cell sensors, both in the presence and absence of the cognate AHL; however, the similarly structured tinidazole and metronidazole activated the pSB1075 sensor to a greater extent than the pSB406 system. It is important to note that when the authors performed the studies with the whole-cell biosensor, they normalized all of the results to cell viability. This was to ensure that the changes observed were due to the interaction of the compound with the regulatory protein and not other interactions with the other components in the cells or due to a decrease in signal due to a decrease in the number of live cells. Another important result to note is that the concentrations of antibiotic that initiated an agonistic response were clinically relevant and included the peak plasma concentrations of the antibiotics.

The authors point out that their results are in contrast to previous studies reported for ciprofloxacin [50]. However, the previous studies were performed in *P. aeruginosa* PAO1 cells whereas the current study was performed using *E. coli* cells. This is an important demonstration of how whole-cell biosensors can allow studying the effects of molecules on a specific protein/promoter system, and in the native cells the effect could be due to other cell components. This was proven by a study showing that azithromycin did not directly affect the expression of QS-related genes, but decreased AHL production by lowering expression of genes upstream to the synthase ones [45].

Whole-cell biosensors are powerful tools, specifically for quorum-sensing regulatory proteins that are difficult to purify and elucidate the structure. The ligand binding domain (LBD) of LasR was isolated and a crystal structure was solved [51]. A study was performed to screen antibiotics for inhibitory QS activity. Specifically, azithromycin and ciprofloxacin were docked against the LBD of LasR. It was shown that these antibiotics do not bind to the AHL binding site of LasR but rather to a different site [50]. By using a whole-cell biosensor, it was shown that although these molecules may not be binding to the LBD, they still bind in such a way that they can cause a conformational change and allow for LasR binding to its promoter region.

4 Conclusions and Future Perspectives

QS is communication between bacteria that allows for the regulation of genes which control certain behaviors such as formation of biofilms, expression of virulence factors, allows for mobility, and bioluminescence, to name a few. This intricate network of cellular communication is dictated by signaling molecules and their associated regulatory proteins. Although there have been continued efforts to develop QSM sensors using regulatory proteins [52], due to the difficulty to express and purify these regulatory proteins, whole-cell biosensors are important to assist in the understanding of the mechanism of QS. Many QS bacteria are implicated in diseases, therefore the studies we have discussed in this chapter have been performed to understand the QS and how it is involved in the disease state. Diagnosis and treatment of a disease are paramount, however, there are instances that could potentially be prevented altogether with proper methods of detection: annually, approximately 48 million people in the United States experience a foodborne illness, resulting in approximately 3000 deaths and an annual cost of illness estimated between US$51–77 billion [53]. Many of the agents involved in causing food borne illness are bacteria [54], with multiple studies having linked QSMs to food spoilage and activation of virulence factors [55]. By focusing on development and optimization of bacterial whole-cell-based biosensing systems for QSMs in foods, it may prove possible to prevent many cases of illness, as well as provide better diagnosis and treatment of severe cases.

Interspecies communication is a growing field of investigation as it plays an important role in large microbial communities. Work has been completed to detect AHLs in ecosystems where a range of bacteria would be present, such as in soil and the rhizosphere [56, 57]. As previously described, certain autoinducer molecules, such as AI-2, are used for interspecies communication. However, continued investigation of QSMs and QS networks is necessary as some organisms such as *Salmonella, Shigella* and *E. coli* cannot synthesize their own AHLs, yet these bacteria are able to communicate with other bacteria. It is hypothesized that this communication is facilitated via a receptor-like protein (SdiA) with amino acid sequences that are homologous to LuxR-type activators [58, 59]. The strain of *E. coli, EHEC*, which results in acute hemorrhagic diarrhea and can cause hemolytic uremic syndrome, appears to produce a thus far unidentified autoinducer termed AI-3. Clarke et al. identified the *E. coli* protein QseC that directly binds both AI-3 and epinephrine/norepinephrine [60]. This was one of the first studies demonstrating that mammalian signaling hormones can be detected by bacteria. AI-3 is thought to activate genes that are involved in intestinal colonization. Other organisms that produce AI-3 are *Shigella, Salmonella, Klebsiella, Enterobacter* and *Citrobacter*. Little is known about how these molecules interact with various quorum-sensing systems, thus whole-cell biosensors could lend insight about which regulatory proteins are important for this interspecies communication and the mechanism of action of these molecules.

Furthermore, QS has been implicated in interkingdom communication. In a novel 3-D co-culture model of epithelial cells and immune cells (monocytes/macrophages) it could be shown that the epithelial cells degrade the *P. aeruginosa* QSM 3-oxo-C12-HSL. This protective effect by the epithelial cells is thought to be caused through enzymatic degradation of these QSMs by the epithelial cells [61]. This study demonstrated that QSMs are sensed not only by bacteria, but also by eukaryocytes. Additionally, QSMs have been shown to cause an inflammatory response [8], which make them an attractive target for therapeutic interventions. Thus by continuing to employ bacterial whole-cell-based biosensors, research may be able to elucidate further how our bodies interact with our biome and affect disease.

Research employing bacterial whole-cell biosensors has proved great insight into the quorum-sensing mechanism. With these tools, it is possible to evaluate molecules and determine which chemical characteristics are important to agonize or antagonize quorum-sensing systems. This information will strongly impact the future of treatment for many of the diseases where QS is implicated, as well as broaden our knowledge of bacterial communication.

Acknowledgments This work was supported in part by grants from the National Science Foundation, the Broad Foundation, Broad Medical Research Program, National Institute of Hometown Security, the Children's Miracle Network, and the Department of Defense. S.D. is grateful for support from the Lucille P. Markey Chair in Biochemistry and Molecular Biology of the Miller School of Medicine of the University of Miami.

References

1. Uroz S, Dessaux Y, Oger P (2009) Quorum sensing and quorum quenching: the yin and yang of bacterial communication. Chembiochem: Eur J Chem Biol 10(2):205–216. doi:10.1002/cbic.200800521
2. Gonzalez JE, Keshavan ND (2006) Messing with bacterial quorum sensing. Microbiol Mol Biol Rev: MMBR 70(4):859–875. doi:10.1128/MMBR.00002-06
3. Waters CM, Bassler BL (2005) Quorum sensing: cell-to-cell communication in bacteria. Annu Rev Cell Dev Biol 21:319–346. doi:10.1146/annurev.cellbio.21.012704.131001
4. Kendall MM, Sperandio V (2014) Cell-to-cell signaling in and. EcoSal Plus 2014. doi:10.1128/ecosalplus.ESP-0002-2013
5. Sifri CD (2008) Healthcare epidemiology: quorum sensing: bacteria talk sense. Clin Infect Dis: Off Publ Infect Diseases Soc Am 47(8):1070–1076. doi:10.1086/592072
6. Sanchez CJ Jr, Mende K, Beckius ML, Akers KS, Romano DR, Wenke JC, Murray CK (2013) Biofilm formation by clinical isolates and the implications in chronic infections. BMC Infect Dis 13:47. doi:10.1186/1471-2334-13-47
7. Raut N, O'Connor G, Pasini P, Daunert S (2012) Engineered cells as biosensing systems in biomedical analysis. Anal Bioanal Chem 402(10):3147–3159. doi:10.1007/s00216-012-5756-6
8. Cohen TP, D Prince A (2015) Host Immune Evasion. Pseudomonas 3–23. doi:10.1007/978-94-017-9555-5_1
9. Rumbaugh KP, Hamood AN, Griswold JA (2004) Cytokine induction by the P. aeruginosa quorum sensing system during thermal injury. The Journal of surgical research 116(1):137–144
10. Gjodsbol K, Christensen JJ, Karlsmark T, Jorgensen B, Klein BM, Krogfelt KA (2006) Multiple bacterial species reside in chronic wounds: a longitudinal study. Int Wound J 3(3):225–231. doi:10.1111/j.1742-481X.2006.00159.x
11. Kirketerp-Moller K, Jensen PO, Fazli M, Madsen KG, Pedersen J, Moser C, Tolker-Nielsen T, Hoiby N, Givskov M, Bjarnsholt T (2008) Distribution, organization, and ecology of bacteria in chronic wounds. J Clin Microbiol 46(8):2717–2722. doi:10.1128/JCM.00501-08
12. Dalton T, Dowd SE, Wolcott RD, Sun Y, Watters C, Griswold JA, Rumbaugh KP (2011) An in vivo polymicrobial biofilm wound infection model to study interspecies interactions. PLoS ONE 6(11):e27317. doi:10.1371/journal.pone.0027317
13. Fekete A, Frommberger M, Rothballer M, Li X, Englmann M, Fekete J, Hartmann A, Eberl L, Schmitt-Kopplin P (2007) Identification of bacterial N-acylhomoserine lactones (AHLs) with a combination of ultra-performance liquid chromatography (UPLC), ultra-high-resolution mass spectrometry, and in-situ biosensors. Anal Bioanal Chem 387(2):455–467. doi:10.1007/s00216-006-0970-8
14. Struss AK, Nunes A, Waalen J, Lowery CA, Pullanikat P, Denery JR, Conrad DJ, Kaufmann GF, Janda KD (2013) Toward implementation of quorum sensing autoinducers as biomarkers for infectious disease states. Anal Chem 85(6):3355–3362. doi:10.1021/ac400032a
15. Kumari A, Pasini P, Daunert S (2008) Detection of bacterial quorum sensing N-acyl homoserine lactones in clinical samples. Anal Bioanal Chem 391(5):1619–1627. doi:10.1007/s00216-008-2002-3
16. Campagna SR, Gooding JR, May AL (2009) Direct quantitation of the quorum sensing signal, autoinducer-2, in clinically relevant samples by liquid chromatography-tandem mass spectrometry. Anal Chem 81(15):6374–6381. doi:10.1021/ac900824j

17. May AL, Eisenhauer ME, Coulston KS, Campagna SR (2012) Detection and quantitation of bacterial acylhomoserine lactone quorum sensing molecules via liquid chromatography-isotope dilution tandem mass spectrometry. Anal Chem 84(3):1243–1252. doi:10.1021/ac202636d
18. Ciofu O, Hansen CR, Hoiby N (2013) Respiratory bacterial infections in cystic fibrosis. Curr Opin Pulm Med 19(3):251–258. doi:10.1097/MCP.0b013e32835f1afc
19. Roy V, Meyer MT, Smith JA, Gamby S, Sintim HO, Ghodssi R, Bentley WE (2013) AI-2 analogs and antibiotics: a synergistic approach to reduce bacterial biofilms. Appl Microbiol Biotechnol 97(6):2627–2638. doi:10.1007/s00253-012-4404-6
20. Singh PK, Schaefer AL, Parsek MR, Moninger TO, Welsh MJ, Greenberg EP (2000) Quorum-sensing signals indicate that cystic fibrosis lungs are infected with bacterial biofilms. Nature 407(6805):762–764. doi:10.1038/35037627
21. Bjarnsholt T (2013) The role of bacterial biofilms in chronic infections. APMIS Suppl 136:1–58. doi:10.1111/apm.12099
22. Middleton B, Rodgers HC, Camara M, Knox AJ, Williams P, Hardman A (2002) Direct detection of N-acylhomoserine lactones in cystic fibrosis sputum. FEMS Microbiol Lett 207 (1):1–7
23. Erickson DL, Endersby R, Kirkham A, Stuber K, Vollman DD, Rabin HR, Mitchell I, Storey DG (2002) Pseudomonas aeruginosa quorum-sensing systems may control virulence factor expression in the lungs of patients with cystic fibrosis. Infect Immun 70(4):1783–1790
24. Chambers CE, Visser MB, Schwab U, Sokol PA (2005) Identification of N-acylhomoserine lactones in mucopurulent respiratory secretions from cystic fibrosis patients. FEMS Microbiol Lett 244(2):297–304. doi:10.1016/j.femsle.2005.01.055
25. Sartor RB (2004) Therapeutic manipulation of the enteric microflora in inflammatory bowel diseases: antibiotics, probiotics, and prebiotics. Gastroenterology 126(6):1620–1633
26. Kumari A, Pasini P, Deo SK, Flomenhoft D, Shashidhar H, Daunert S (2006) Biosensing systems for the detection of bacterial quorum signaling molecules. Anal Chem 78 (22):7603–7609. doi:10.1021/ac061421n
27. Winson MK, Swift S, Fish L, Throup JP, Jorgensen F, Chhabra SR, Bycroft BW, Williams P, Stewart GS (1998) Construction and analysis of luxCDABE-based plasmid sensors for investigating N-acyl homoserine lactone-mediated quorum sensing. FEMS Microbiol Lett 163 (2):185–192
28. Yates EA, Philipp B, Buckley C, Atkinson S, Chhabra SR, Sockett RE, Goldner M, Dessaux Y, Camara M, Smith H, Williams P (2002) N-acylhomoserine lactones undergo lactonolysis in a pH-, temperature-, and acyl chain length-dependent manner during growth of Yersinia pseudotuberculosis and Pseudomonas aeruginosa. Infect Immun 70(10):5635–5646
29. Yang F, Wang LH, Wang J, Dong YH, Hu JY, Zhang LH (2005) Quorum quenching enzyme activity is widely conserved in the sera of mammalian species. FEBS Lett 579(17):3713–3717. doi:10.1016/j.febslet.2005.05.060
30. Raut N, Pasini P, Daunert S (2013) Deciphering bacterial universal language by detecting the quorum sensing signal, autoinducer-2, with a whole-cell sensing system. Anal Chem 85 (20):9604–9609. doi:10.1021/ac401776k
31. Kumar R, Chhibber S, Gupta V, Harjai K (2011) Screening & profiling of quorum sensing signal molecules in Pseudomonas aeruginosa isolates from catheterized urinary tract infection patients. Ind J Med Res 134:208–213
32. Senturk S, Ulusoy S, Bosgelmez-Tinaz G, Yagci A (2012) Quorum sensing and virulence of Pseudomonas aeruginosa during urinary tract infections. J Infect Dev Countries 6(6):501–507
33. Lakshmana Gowda K, John J, Marie MA, Sangeetha G, Bindurani SR (2013) Isolation and characterization of quorum-sensing signalling molecules in Pseudomonas aeruginosa isolates recovered from nosocomial infections. APMIS: Acta Pathol Microbiol Immunol Scand. doi:10.1111/apm.12047
34. Gupta P, Gupta RK, Harjai K (2013) Quorum sensing signal molecules produced by Pseudomonas aeruginosa cause inflammation and escape host factors in murine model of urinary tract infection. Inflammation 36(5):1153–1159. doi:10.1007/s10753-013-9650-y

35. Daunert S, Barrett G, Feliciano JS, Shetty RS, Shrestha S, Smith-Spencer W (2000) Genetically engineered whole-cell sensing systems: coupling biological recognition with reporter genes. Chem Rev 100(7):2705–2738
36. Feliciano JP, Deo P, Daunert, S (2006) Photoproteins as reporters in whole-cell sensing. Wiley-VCH Verlag GmbH & Co KGaA 131–154
37. Yong YC, Zhong JJ (2009) A genetically engineered whole-cell pigment-based bacterial biosensing system for quantification of N-butyryl homoserine lactone quorum sensing signal. Biosens Bioelectron 25(1):41–47. doi:10.1016/j.bios.2009.06.010
38. Struss A, Pasini P, Ensor CM, Raut N, Daunert S (2010) Paper strip whole cell biosensors: a portable test for the semiquantitative detection of bacterial quorum signaling molecules. Anal Chem 82(11):4457–4463. doi:10.1021/ac100231a
39. Miller MB, Bassler BL (2001) Quorum sensing in bacteria. Annu Rev Microbiol 55:165–199. doi:10.1146/annurev.micro.55.1.165
40. Federle MJ (2009) Autoinducer-2-based chemical communication in bacteria: complexities of interspecies signaling. Contrib Microbiol 16:18–32. doi:10.1159/000219371
41. Manichanh C, Rigottier-Gois L, Bonnaud E, Gloux K, Pelletier E, Frangeul L, Nalin R, Jarrin C, Chardon P, Marteau P, Roca J, Dore J (2006) Reduced diversity of faecal microbiota in Crohn's disease revealed by a metagenomic approach. Gut 55(2):205–211. doi:10.1136/gut.2005.073817
42. Baumgart M, Dogan B, Rishniw M, Weitzman G, Bosworth B, Yantiss R, Orsi RH, Wiedmann M, McDonough P, Kim SG, Berg D, Schukken Y, Scherl E, Simpson KW (2007) Culture independent analysis of ileal mucosa reveals a selective increase in invasive Escherichia coli of novel phylogeny relative to depletion of Clostridiales in Crohn's disease involving the ileum. ISME J 1(5):403–418. doi:10.1038/ismej.2007.52
43. Swift S, Throup JP, Williams P, Salmond GP, Stewart GS (1996) Quorum sensing: a population-density component in the determination of bacterial phenotype. Trends Biochem Sci 21(6):214–219
44. Reverchon S, Chantegrel B, Deshayes C, Doutheau A, Cotte-Pattat N (2002) New synthetic analogues of N-acyl homoserine lactones as agonists or antagonists of transcriptional regulators involved in bacterial quorum sensing. Bioorg Med Chem Lett 12(8):1153–1157
45. Kai T, Tateda K, Kimura S, Ishii Y, Ito H, Yoshida H, Kimura T, Yamaguchi K (2009) A low concentration of azithromycin inhibits the mRNA expression of N-acyl homoserine lactone synthesis enzymes, upstream of lasI or rhlI, in Pseudomonas aeruginosa. Pulm Pharmacol Ther 22(6):483–486. doi:10.1016/j.pupt.2009.04.004
46. Struss AK, Pasini P, Flomenhoft D, Shashidhar H, Daunert S (2012) Investigating the effect of antibiotics on quorum sensing with whole-cell biosensing systems. Anal Bioanal Chem 402 (10):3227–3236. doi:10.1007/s00216-012-5710-7
47. Gillis RJ, Iglewski BH (2004) Azithromycin retards Pseudomonas aeruginosa biofilm formation. J Clin Microbiol 42(12):5842–5845. doi:10.1128/JCM.42.12.5842-5845.2004
48. Nalca Y, Jansch L, Bredenbruch F, Geffers, R, Buer J, Haussler S (2006) Quorum-Sensing Antagonistic Activities of Azithromycin in Pseudomonas aeruginosa PAO1: a Global Approach. Antimicrob. Agents Chemother 50:1680–1668
49. Bala A, Kumar R, Harjai K (2011) Inhibition of quorum sensing in Pseudomonas aeruginosa by azithromycin and its effectiveness in urinary tract infections. J Med Microbiol 60(Pt 3):300–306. doi:10.1099/jmm.0.025387-0
50. Skindersoe ME, Alhede M, Phipps R, Yang L, Jensen PO, Rasmussen TB, Bjarnsholt T, Tolker-Nielsen T, Hoiby N, Givskov M (2008) Effects of antibiotics on quorum sensing in Pseudomonas aeruginosa. Antimicrob Agents Chemother 52(10):3648–3663. doi:10.1128/AAC.01230-07
51. Bottomley MJ, Muraglia E, Bazzo R, Carfi A (2007) Molecular insights into quorum sensing in the human pathogen Pseudomonas aeruginosa from the structure of the virulence regulator LasR bound to its autoinducer. J Biol Chem 282(18):13592–13600. doi:10.1074/jbc.M700556200

52. Raut N, Joel S, Pasini P, Daunert S (2015) Bacterial autoinducer-2 detection via an engineered quorum sensing protein. Anal Chem 87(5):2608–2614. doi:10.1021/ac504172f
53. Scharff RL (2012) Economic burden from health losses due to foodborne illness in the United States. J Food Prot 75(1):123–131. doi:10.4315/0362-028X.JFP-11-058
54. Scallan E, Griffin PM, Angulo FJ, Tauxe RV, Hoekstra RM (2011) Foodborne illness acquired in the United States–unspecified agents. Emerg Infect Dis 17(1):16–22. doi:10.3201/eid1701.091101p2
55. Skandamis PN, Nychas GJ (2012) Quorum sensing in the context of food microbiology. Appl Environ Microbiol 78(16):5473–5482. doi:10.1128/AEM.00468-12
56. DeAngelis KM, Firestone MK, Lindow SE (2007) Sensitive whole-cell biosensor suitable for detecting a variety of N-acyl homoserine lactones in intact rhizosphere microbial communities. Appl Environ Microbiol 73(11):3724–3727. doi:10.1128/AEM.02187-06
57. Burmolle M, Hansen LH, Sorensen SJ (2005) Use of a whole-cell biosensor and flow cytometry to detect AHL production by an indigenous soil community during decomposition of litter. Microb Ecol 50(2):221–229. doi:10.1007/s00248-004-0113-8
58. Michael B, Smith JN, Swift S, Heffron F, Ahmer BM (2001) SdiA of Salmonella enterica is a LuxR homolog that detects mixed microbial communities. J Bacteriol 183(19):5733–5742. doi:10.1128/JB.183.19.5733-5742.2001
59. Smith JN, Ahmer BM (2003) Detection of other microbial species by Salmonella: expression of the SdiA regulon. J Bacteriol 185(4):1357–1366
60. Clarke MB, Hughes DT, Zhu C, Boedeker EC, Sperandio V (2006) The QseC sensor kinase: a bacterial adrenergic receptor. Proc Natl Acad Sci USA 103(27):10420–10425. doi:10.1073/pnas.0604343103
61. Crabbe A, Sarker SF, Van Houdt R, Ott CM, Leys N, Cornelis P, Nickerson CA (2011) Alveolar epithelium protects macrophages from quorum sensing-induced cytotoxicity in a three-dimensional co-culture model. Cell Microbiol 13(3):469–481. doi:10.1111/j.1462-5822.2010.01548.x

Adv Biochem Eng Biotechnol (2016) 154: 201–204
DOI 10.1007/978-3-319-27407-2

Index

MIX
Papier aus verantwortungsvollen Quellen
Paper from responsible sources
FSC® C105338

If you have any concerns about our products,
you can contact us on
ProductSafety@springernature.com

In case Publisher is established outside the EU,
the EU authorized representative is:
Springer Nature Customer Service Center GmbH
Europaplatz 3, 69115 Heidelberg, Germany

Printed by Libri Plureos GmbH
in Hamburg, Germany